MANAGING FOR
WORLD CLASS SAFETY

MANAGING FOR WORLD CLASS SAFETY

J. M. STEWART

A Wiley-Interscience Publication

JOHN WILEY & SONS, INC.

For ordering and customer service, call 1-800-CALL-WILEY.

Library of Congress Cataloging-in-Publication Data:
Stewart, J. M.
    Managing for world class safety / J. M. Stewart.
      p. cm.
    "A Wiley-Interscience publication".
    Includes bibliographical references.
    ISBN 0-471-44386-7 (cloth : alk. paper)
    1. Industrial satety—Management. 2. Psychology, Industrial. I. Title
HD7262 .S813 2002
658.3′82—dc21

                2001037875

Printed in the United States of America.

10 9 8 7 6 5 4

CONTENTS

FOREWORD

The safety of its people has always been the most important value at DuPont. It shows in the enduring world-class safety record of the company. Those of us who have worked for such companies and have become steeped in the philosophy of safety management, often tend to forget how difficult it is to manage safety to high levels of excellence. It requires a strong commitment to fundamental values and thorough application of the key practices. For example, the central belief that *all* injuries can be prevented must become more than a slogan. It must be the driver that requires all leaders to work persistently to prevent any injury—and to take full responsibility if one does occur.

While there have been many articles and books that describe how the safest companies achieve excellence, in most cases the evidence is qualitative and anecdotal. It could be said that safety management has not advanced to as sophisticated a level as other areas of management. There has been an evident need for more systematic and quantitative research on safety management, done in such a way that companies seeking to improve could find more specific guidance. Part of the problem has been that the foundations of safety management lie in a system of beliefs—a culture of safety—and measurement in these "soft" areas is difficult.

It is thus very gratifying to see the results of Jim Stewart's research. He has thoroughly documented the basic beliefs and practices of safety management and then developed methods to "measure" the extent they are present in an organization. The benchmark data obtained through careful research at some of the safest companies and some with very poor safety represents an important step in improving the understanding of safety management. Jim's techniques were originally developed for use in safety consulting and have been widely proven in practice with Canadian companies. The combination of research and consulting ensured that the research was soundly based on best practice and useable and credible with corporate leaders.

Jim Stewart worked closely with me for many years in senior positions when I was CEO of DuPont Canada. He was always an innovator, interested in finding out how to make management systems work better and to develop and introduce new ones. And then to find ways to conceptualize them so that they could be applied elsewhere. He was one of the leaders in building self-management and world class manufacturing into the company, a legacy that endures today. He could envisage a better way, test it methodically and drive it to practice across the company. In the 1980s, Jim was in charge of the company's manufacturing and engineering as the company moved to self-management systems and adapted to global competition. Through this period of great change the company maintained and improved its safety performance. Jim was the first recipient of the company's highest award—the Daedalus Award—given for his outstanding contribution to building self-management, increasing productivity and reducing costs.

Thus it did not surprise me that when Jim moved into the field of research and consulting on the management of safety, he was able to develop new, leading-edge systems. His background in research, in management and his involvement in safety were ideal qualifications for this.

The research described in this book was built on the solid principles of safety management practiced by the safest companies. It breaks new ground in developing techniques to *measure* the state of safety management quantitatively. It advances the knowledge of safety management through the investigations carefully done at very safe companies and at companies with very poor safety. Thus it lays the base for a much more specific methods for companies to assess where they stand in comparison to best practice, and to focus directly where they should seek to improve.

It is a very substantial and very important undertaking. For those organizations that wish to make major changes in their safety performance it will be very valuable. It should be compulsory reading for every member of every leadership team.

Ted Newall
Chairman, Nova Chemical Corporation
Former Chairman, President & CEO, DuPont Canada Inc.

PREFACE

In the world's safest companies, safety has unquestioned priority and meticulous attention is given to using the best safety practices. As a long-term executive at DuPont, one of the very safest companies, I had known that such attention to safety was the exception, not the rule. But it was only when I became involved in consulting on safety management that I realized just how formidable the obstacles are for leaders intent on making a step change in their company's safety performance. The main barriers are:

1. There are few good techniques to "measure" the state of safety management, in particular to measure the intangibles such as management commitment. Thus assessment of safety is usually observational and anecdotal rather than quantitative.
2. Although there are descriptions of how the safest companies manage safety, because of the lack of measurement tools, there and few quantitative benchmark data. There is poor understanding of what constitutes "World Class Safety," particularly understanding of the central role played by specific safety values.
3. Partly because of the inadequacy of the assessment tools and the lack of benchmark data, management is often reluctant to undertake the fundamental changes required to reform safety. Better ways are needed to help convince them that a step change can be managed through orderly processes.

As a result of these interrelated factors, improvement efforts often do not focus on the "right things." The right things are usually not the physical or system deficiencies that are the easiest to see. Rather, they are the intangibles that cannot be easily measured and thus are often neglected—like lack of management commitment or a low level of involvement of workers in safety activities.

In my consulting practice, before undertaking this research, I had developed a Safety Survey process to deal more thoroughly and quantitatively with the first of these factors—the measurement of the state of safety management. The process was built on a model of safety management supported by a set of beliefs and practices derived from the methods of very safe companies. The central innovation of the Safety Survey was a comprehensive safety questionnaire, completed by a cross section of the organization, from leaders to workers. The questionnaire "measures" the extent to which the factors of the model and the beliefs and practices are present and effective in the workplace.

The research described in this book addresses the second of the above three issues—the lack of comprehensive information, particularly quantitative benchmark data, on how the safest companies achieve excellence. The research was undertaken from 1995 through 1998 while I was an Executive-in-Residence at the Rotman School of Management at the University of Toronto. The Safety Survey techniques were used to investigate in depth five North American companies that have achieved enduring world class safety performance and to determine, in a more quantitative way than had previously been done, how they achieve excellence. Similar measurements were completed at five companies with very poor safety, providing a striking contrast to the safe company data and helping to validate the measurement techniques.

The lost work injury frequency of the five very safe companies averaged 0.08 per 200,000 hours over a 5-year period. The 24-question, multiple-choice questionnaire was completed by 400 people at the five companies. They ranged from workers to presidents. Extensive on-site research was conducted with the five very safe companies to supplement the questionnaire results.

At the other end of the safety performance scale, 250 employees of five companies with very poor safety records completed the same questionnaire. The lost work injury frequency of these companies averaged 20 per 200,000 hours, 250 times worse than that of the very safe companies. The surveys of the companies with very poor safety were conducted with the help of provincial safety agencies. On-site investigations beyond the completion of the questionnaires were not possible.

For almost every question, the quantitative results of the research correlated well with safety performance. In most cases there were dramatic differences between the results from the very safe companies and those from the companies with very poor safety. For example, 75% of the 400 respondents in the safe companies reported that their management was held responsible for injuries, compared with only 16% of the 250 respondents from the companies with poor safety. The contrast between the "best" results from the safe companies and the "worst" results from the companies with poor safety was even more striking. For example, in one of the safe companies, 68% of workers said that they were deeply or quite involved in safety activities and a further 20% said that they were moderately involved. In the worst result from an unsafe company, only 12% of workers said they were deeply or quite involved; 88% said they were not much involved or not involved at all.

Through the research, new insights were gained into the fundamentals of safety management. The model of safety management and the questionnaire were improved and extended. To augment the survey data and to illustrate how they achieve outstanding safety, a case write-up was prepared on each of the five very safe companies.

The results of the research confirm the validity of the model of safety management. The most important factors driving excellence in safety prove to be management commitment, line ownership of safety, and worker involvement. If these "soft" factors are present, they in turn lead to excellence in the important safety practices.

The research further confirmed the power of the questionnaire technique for measuring the state of safety management. It affords a way to quantitatively assess the important intangibles such as management commitment and to identify the quality and effectiveness of a company's safety practices. By determining in specific terms where the problems lie, the survey helps indicate the pathway to improvement. The quantitative results developed in this research represent unique benchmark data, invaluable in helping assess the state of safety management in a company.

A research report on the work was published in 1999, and this book closely follows that report. The copyrighted elements of the Safety Survey process, including the questionnaire and associated implementation techniques, initially developed by the author's consulting company, were licensed to E. I. DuPont de Nemours in 2000 and are being integrated into its safety consulting business.

J. M. Stewart

ACKNOWLEDGMENTS

The research described in this book was carried out at the Rotman School of Management at the University of Toronto while the author was located there as an Executive-in-Residence and Adjunct Professor of Strategic Studies. The author thanks the School and the University for providing the facilities and for assistance in the research. Very capable help was provided by Kazuo Noguchi, a research assistant at the School, and later by Peter Steer.

The research was funded by donations to the University from Canadian governments and Canadian corporations. All of the provincial and territorial safety or workers' compensation agencies and the federal Departments of Health and Human Resources participated. The government funding was coordinated by Human Resources Development Canada (HRDC) through the Occupational Safety and Health Committee of the Canadian Association of Administrators of Labour Legislation (CAALL-OSH Committee). The author is grateful to HRDC, to the members of the CAALL-OSH Committee, and to their respective governments for their advice as well as for their financial support. The late Jim McLellan, formerly of HRDC, was instrumental in initiating the research and was a strong supporter of the work.

The companies participating in the funding were Dow Chemical Canada, Nacan Products, National Rubber Co., Nova Chemicals, Ontario Hydro, Shell Canada, and TransCanada Pipelines. The author thanks these companies sincerely for their generous contributions.

A central part of the project was the detailed research done at five companies with excellent safety records—Abitibi-Consolidated, DuPont Canada, Milliken and Company (US), S&C Electric Canada, and Shell Canada. These companies went the second mile in many ways, opening their facilities for the author, making the time of their people freely available, and providing the information needed for the research. The author thanks these companies and their people for their outstanding cooperation.

The collection of questionnaire data at companies with very poor safety was conducted by the provincial agencies of British Columbia (WCB of B.C.), Manitoba (Department of Labour), and Ontario (Industrial Accident Prevention Association—IAPA). This work could only be done through such public bodies, and their generous help provided invaluable data for the project.

The draft of the research report on which this book was based was reviewed by James Hansen of the IAPA, John Shepherd, formerly of Nacan, Allan Luck of the B.C. WCB, and Linton Kulak and Ron Czura of Shell Canada. The author thanks them sincerely for their very helpful comments.

Throughout the research, particular help was provided by Maureen Shaw and others at the IAPA (Ontario), by Geoff Bawden and Barry Warrack of the Manitoba Department of Labour, and by Allan Luck and Ralph McGinn at the British Columbia WCB.

1

INTRODUCTION

Every year, over 700 Canadians are killed in workplace accidents and 400,000 are injured seriously enough to require time off the job (1). In addition to the toll in human suffering, occupational injuries and illnesses cost the Canadian economy tens of billions of dollars per year. If all the indirect costs are added up, the tally probably comes to more than $25 billion per year (2). Although the injury frequency has been declining, costs have increased substantially. This situation exists in the face of a surprising observation: the safest companies achieve results up to 1000 times better than the worst and 10 to 100 times better than the average! How can such differences be justified? The know-how to achieve much better safety performance apparently exists. Why is it not being applied?

The costs of injuries are sometimes compared with the costs of safety initiatives in the context of what is called the cost-benefit trade-off. The assumption is that beyond achieving fair safety, further efforts to improve safety would be counterproductive. A major reason for poor safety performance is that many leaders work from this premise.

Not much evidence has been published to support the ability of firms to achieve world class safety and, at the same time, excellent business performance. Two papers on this subject by the author were behind the initiation of this research project. The first paper described the deterioration in safety at DuPont Canada in the late 1980s and the subsequent recovery back to outstanding safety at the same time as profitability was improved to high levels (3). The second paper reported a tenfold improvement in safety at National Rubber along with a turnaround from loss to profit (4). These papers stressed that excellence in safety can be achieved through a management-by-principle approach. They emphasized the critical requirement that corporate leaders, particularly CEOs, become committed to safety excellence. Both papers drew the analogy to quality—safety must be built into the way everyone does the job and does it right the first time. The papers were largely anecdotal, as are most similar references in the literature.

Jim McLellan, the Director General of Health and Safety at Human Resources Development Canada in the mid-1990s, observed that there were few articles in the literature that described Canadian cases. He proposed that research be undertaken that would help convince business leaders to take up the cause of improvement in safety.[i] Funding was obtained from Canadian governments and corporations, and the research was started in 1995.

The original objective of the project was to seek examples of companies with both excellent safety *and* outstanding business records, and thus the project was originally entitled "World Class Safety and Outstanding Business Performance." The intent was not to try to correlate safety and business performance quantitatively, but it was envisaged that in selecting companies for study, some sort of "business success" criteria should be met. However, as the research proceeded, the most difficult problem proved to be finding companies that met world class *safety* standards. Those that did meet the safety standards proved to be leading companies in business performance. Thus no specific criteria for assessing business success were developed. The project became much more an attempt to find out quantitatively and in depth how a few companies with outstanding safety managed to achieve and maintain such excellence.

The original goal was to study 7 to 10 companies, mainly in Canada, but also including Asian, US, and European examples. Such broad geographic coverage proved to be beyond the resources of the project, and thus this is largely a North American study. It assessed the management of safety in five very safe companies—one US and four Canadian. The study could not realistically cover the whole range of industry sectors, so the research was confined to goods-producing companies in the manufacturing and natural resource industries.

---

[i] The research dealt with safety rather than the broader subject of occupational health and safety.

In the areas mentioned above, the research was restricted to less than the original intent. However, in other ways its scope was expanded and deepened.

Before undertaking this research, the author had developed a Safety Survey process to assist in consulting projects on safety management. The process was built on a model of safety management supported by a set of beliefs (values)[ii] and practices derived from the methods of very safe companies. The central innovation of the Safety Survey process was a comprehensive safety questionnaire, completed by a cross section of the company's workforce, from leaders to workers. The questionnaire "measures" the extent to which the factors of the model and the beliefs and practices are present and effective in the workplace. In consulting, the Safety Survey process had proven to be a unique and powerful tool for diagnosing the state of safety management and providing guidance for improvement. It also proved to be an ideal tool for research and was used to develop quantitative data on the methods and practices of the five very safe companies.

As the research planning progressed, it became evident that the survey process could be further "calibrated" by applying it at companies with *very poor* safety as well as at companies with world class safety. Thus the survey was extended to cover five companies with very poor safety performance in addition to the five with excellent safety.

The research was completed in 1998, and the research report on which this book is based was issued in June 1999 (5). The book essentially follows the report, with some editorial changes and corrections of minor errors in the data.

As the research proceeded, the overall objectives were modified to incorporate these changes in scope. The final objectives are given below.

Objectives of the Research Project

1. To further develop the model of managing for outstanding safety that generalizes the concepts, beliefs, and practices used by very safe companies in such a way that the knowledge can be transferred to other organizations. *(Specific Product: A Model of Managing for Outstanding Safety)*

2. To further develop the safety questionnaire as a quantitative method for assessing the state of safety management in a company—a means of *measuring* the extent that the concepts,

---

[ii] Throughout this book, *beliefs* will be used as a surrogate for *values.*

beliefs, and practices of the model are present in an organization. (*Specific Product: A Safety Questionnaire*)

3. To conduct detailed surveys at five companies with outstanding safety records and five companies with very poor safety. This would validate and help improve the model and the questionnaire and provide benchmark data. (*Specific Product: quantitative questionnaire data from five very safe companies and from five with very poor safety*)

4. To support the quantitative data from the questionnaires at each of the five very safe companies with on-site investigations, interviews, and focus groups (not done for the companies with poor safety). This would augment the questionnaire data, helping identify the ways that the five companies achieve excellence in safety. (*Specific Products: supporting information to augment the questionnaire data; a case write-up for each of the five very safe companies*)

5. To develop a comprehensive theory of the management of safety and to test its validity through the research findings. (*Specific Products: a report, papers, and presentations that integrate the research findings into a comprehensive theory of safety management*)

Organization of the Book

1. **Introduction**
2. **The Model of Safety Management.** The model of managing for outstanding safety and the beliefs and practices that lie behind it are described.
3. **The Safety Questionnaire.** The background and rationale of the safety questionnaire and its features are described.
4. **Selection of Companies for Research.** The basis for the selection of the companies and a summary of their safety records is presented.
5. **The Research Methodology.** The procedures for conducting the questionnaire surveys, focus groups, and other investigations are described.
6. **Analysis of the Questionnaire Results.** The results from the questionnaire surveys are presented and discussed.
7. **The Safety Management Approaches of Five Very Safe Companies.** The specific ways that each of the five companies manages safety are described, augmenting the questionnaire results with

information from site observations and investigations and from interviews and focus groups.

8. **Conclusions—How Companies Achieve World Class Excellence in Safety.** The overall findings are assessed in terms of the model as a comprehensive theory of safety management. The validity of the model and of the questionnaire is discussed.

9. **Applying the Results of the Research.** The use of the research findings for improving safety is discussed.

The data and a number of subsidiary subjects are included in the Appendices.

The companies that cooperated in the research have provided a great deal of data and allowed their employees to be interviewed, take part in focus groups, and answer detailed questionnaires. The database includes completed questionnaires from about 650 people in 10 companies. With the exception of some examples from the very safe companies, individual company data are not given or, if they are given as examples of the "best" or "worst" results, the company is not identified. The identity of the companies with very poor safety cannot be revealed.

2

THE MODEL OF SAFETY MANAGEMENT

Before this research was undertaken, a model of safety management had been developed by the author. It provided the foundation for a safety questionnaire that had been used successfully in consulting. Validating and further improving the model was an important part of the research project, particularly because the questionnaire became the central tool in conducting the research.

The model provides a framework for understanding and ordering observations of workplace safety, identifying the factors that are needed for excellence, and describing how they are related to each other. It also helps to communicate the concepts of safety management by defining the fundamental drivers of excellence and by providing a road map for planning the action to improve safety. The model was built from observation of companies with outstanding safety—those with records up to 100 times better than the average in indicators such as lost work injury frequency.

2.1 THE FRAMEWORK OF THE MODEL

The relationship among the factors[i] that deliver excellence in safety— the framework of the model—is shown in the schematic diagram in Fig. 2-1.

[i] *Factors* is used to designate the seven main variables in the model that influence safety—the five independent variables of management commitment, line ownership,

6

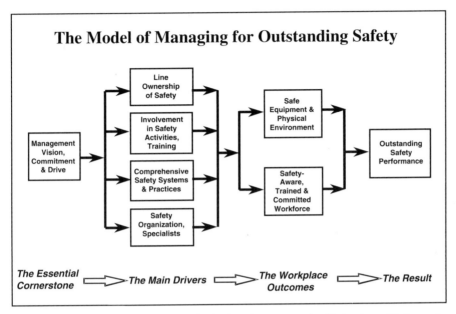

Figure 2-1   In the Model Framework, the key factors are the "behavioural" elements.

The model describes the requirements for *outstanding* safety. It is based on the observation that excellence in safety begins with management commitment, flowing from fundamental beliefs and driven by a vision of excellence. Outstanding safety requires leaders with a "passion" for safety. The vision lays out the desired safety culture, establishes long-term goals, and defines the standards of excellence expected. The vision is implemented through "ownership" of the safety agenda by the line managers.[ii] They take full responsibility for driving safety, for seeing that no one is injured. They are fully accountable for the results. The attitudes and behaviour of the workforce[iii] are aligned to the corporate vision through the demonstrated commitment—the visible action—of all of the line management, led by the CEO. This is

involvement and training, safety systems, and safety organization and the two dependent variables of safe equipment and a safety-aware workforce.

[ii] Line management or "the line," is defined as starting with the CEO and cascading down to and including individual workers. Each manager is fully responsible for the safety of every person reporting to him or her and is accountable for injuries that occur to them. Each person is fully responsible for his or her own safety and less directly for the safety of co-workers. This is similar to the IRS or internal responsibility system.

[iii] "Workforce" means everyone in the workplace—managers, supervisors, and workers.

bolstered by thorough communication of values and goals and by safety training. Above all, attitude and behaviour are influenced by involvement of everyone in the workforce, particularly those on the front line, in "doing things in safety."

Management commitment, line ownership, and workforce involvement are the fundamental "drivers" of safety. These "soft" factors are supported by comprehensive safety systems and practices. In organizations with excellent safety, the systems and practices are meticulously implemented and continuously improved. Their quality and effectiveness are good indicators of management's commitment. Teams drawn from line management and the workforce (broadly) operate the systems and practices, thus promoting "ownership" of and commitment to the values that lie behind the practices.

The roles of the safety organization and of safety specialists are built on the same principles of line ownership and worker involvement. Their task is to facilitate the actions of management, supervision and workers, not to lead the safety agenda. Although the safety organization and safety specialists contribute to excellence, their role is not as fundamental as behavioural and attitudinal factors such as management commitment.

In this model, a safe physical environment and a safety-aware attitude in the workforce are seen as *outcomes* rather than causes of excellence.

The diagram of the relationships among the factors—the framework of the model—is fairly simple. It is a "truism" in that it could be applied to the management of other things such as quality. The safest companies have learned from experience that the factors are the key determinants of excellence. However, none of the factors—the *essential cornerstone*, the *main drivers*, or the *workplace outcomes*—can be easily measured. Only the *results* can be measured directly, in terms of injury frequency, for example.

The difficulty of measurement is one of the main reasons for the ineffectiveness of many assessments of safety, whether carried out internally or by a consulting company or a regulatory agency. Because the soft factors cannot be easily assessed, audits usually concentrate on the factors that *can* be observed directly—the facilities and the safety systems and practices. The assessment of critically important factors such as management commitment, line ownership, and worker involvement depends on subjective observation. Auditors are usually not experienced in senior management and do not appreciate the overriding importance of these factors. They do not realize that excellence in equipment, systems and safety technology is derivative—a *result* of management commitment and the power of the safety culture. Often workplaces that appear to have safe equipment and facilities, good safety systems, and plenty of experts in safety have mediocre safety records.

Even when the importance of the soft factors is realized, there are inadequate tools to assess their presence and influence. Most importantly, it is difficult to make good comparisons to benchmarks. The safe companies say they give first priority to safety, but managers in companies with mediocre safety say the same thing. There is obviously a difference in the way they manage, but what solid evidence can be presented to convince them that their assignment of priority is nothing like that of the very safe companies? How can an auditor assess their priority assignments quantitatively?

The main purpose of this research was to determine more definitively how the safest companies manage to stand so far above the pack and to characterize this excellence quantitatively. There are plenty of *qualitative* descriptions of the way the safest companies manage. Their commitment and their passion for safety is evident in the depth and power of their safety culture and in the excellence of their results. But what are the specifics that characterize this excellence? How can they be *measured* in such a way that their experience can be transferred to other organizations?

2.2   DESIGNING THE MODEL FOR MEASUREMENT

Most of the models of safety management in the literature are descriptive in nature. Although they provide guidance on methods of safety management, they are not constructed to allow quantitative measurement.

Many of the models could be called operational models. They describe the organization of safety management—the central safety committee, the standing subcommittees, the safety specialists, etc. The functions of these bodies and their relationship to the corporate structure and processes are described (see, e.g., Refs. 6 and 7). These models are useful, but they do not lend themselves to even semi-quantitative measurement.

Other models illustrate the connections from policy to principles to programs (see, e.g., Ref. 8). They present a dimension of safety management different from the operational models and complementary to them. Although they are also valid and useful, they do not facilitate measurement, either.

A different, more experimental approach would be to build a model by studying a large number of companies with a range of safety records. The factors hypothesized to be causative would be measured for each company. Statistical methods might then identify the most important determinants of excellence in safety. A technique like this was used successfully by the Strategic Planning Institute to assess the effects of business factors on profitability (9). Data was collected for many years for more than 3000 businesses. The technique would likely work for safety. However, it would call for a massive amount of data. It would require research on organizations with a wide range of safety records that had installed some of the elements of the model but not others. A somewhat similar approach was used by Shannon et al. (10), who postulated a large number of possible variables and then tried to ascertain their relevance to safety management by surveying several hundred Ontario companies. The data collected for each company were limited, and few correlations were found.

Most of the benchmarking studies in the literature are also qualitative. They generally rely on assessing the state of safety management by observation or by subjective questioning. Even if the checklists used for benchmarking are thoroughly built on best practices, the results are still qualitative and subjective.

The intention of the model described here was to provide a way to break through these barriers and provide a foundation for assessing the state of safety management quantitatively. This placed unusual strictures on its design, tougher demands than for the usual descriptive models. The model had to be comprehensive, missing no important variables. The variables had to be measurable, even though many of them were "soft."

Behaviour-based safety systems are increasingly described in the literature (see, e.g., Refs. 11 and 12). Some of them are beginning to be used as a basis for measurement (e.g., Refs. 12–14). They are really subsets of self-management systems, which are principle based and depend on developing beliefs, building commitment to them, and using them to guide the organization's operations. Self-management systems have become the foundation for safety management at some companies. At DuPont Canada, where self-management is well entrenched, a set of safety beliefs has been in place for many years, although not formalized in a way that could be used for measurement. (Refs. 3, 15, and 16 describe some aspects of self-management and the use of safety beliefs at DuPont Canada.)

The framework of the model, itself behaviour based, had been extended to include a comprehensive set of beliefs and practices. This facilitated its use for quantitative measurement of the state of safety management.

The Model of Managing for Outstanding Safety thus includes three layers:

- the *framework* linking the main factors (shown in Fig. 2-1)
- the *values*—the beliefs (and principles) underlying the framework[iv]
- the *practices* that bring the beliefs to life in the workplace

Supporting the complex factors of management commitment, line ownership, and involvement is a set of beliefs and practices. In organizations with enduring excellence in safety, the safety values are deeply held and they are brought to life in the *practices*.

Values (Beliefs) → Attitudes → Behaviour (Practices)

The beliefs and practices are more specific than the factors in the model and thus provide a basis for measuring the state of safety management in an organization and for comparing the results to those from other organizations.[v] Thus an important step in building the model was to define the key beliefs and practices that characterize the culture in the safest companies. Techniques such as questionnaires could be then designed to measure the extent to which they are present in the safest workplaces and to compare the results to those from workplaces with poor safety.

The model of safety management described here differs from many of the models in the literature. It is based primarily on behavioural concepts, but that in itself is not unique. The factors in the framework of the model are well known, although their interrelationship is not usually defined precisely. Similar concepts have been in place for many years at companies such as DuPont, Shell, and others. For example,

---

[iv] Values are defined as the combination of beliefs (what we hold to be true) and principles (the guidelines for translating beliefs into action). In this book, *beliefs*, rather than principles, are discussed. Beliefs are more fundamental and in some cases do not result directly in a single principle. Phrasing values as beliefs forces people to consider their basic thinking on the subject and exposes conflicts and lack of alignment.

[v] There is not a direct one-to-one correspondence among the factors in the framework in Fig. 2-1 and the beliefs and the practices. For example, management commitment to excellence in safety, one of the framework factors, underlies all the beliefs and practices.

a similar model was described as the basis for safety management at Intel (17). Many of the beliefs and most of the practices are in use at these companies, although the list of beliefs is usually not as comprehensive, because the purpose is descriptive and qualitative. What is perhaps unique here is the assembling of the concepts, beliefs, and practices into a comprehensive, ordered model in a way that facilitates measurement.

The validity of the model rests partly on its basis in the experience of a range of companies with outstanding safety. The research results provide an important test. If the model is valid, the safety performance of companies with excellent and poor safety should correlate with the extent to which the factors, beliefs, and practices are present and the thoroughness of their use.

2.3   THE BELIEFS AND PRACTICES FOR EXCELLENCE IN SAFETY

Most of the very safe companies have written safety beliefs (values). The purpose of formalizing them is to provide a comprehensive foundation for providing direction to the workforce. The everyday stressing of values is one of the best ways for leaders to communicate their expectations and to build safety awareness. Although there are different ways of stating the beliefs, there is general agreement on the most important ways.

The beliefs are made explicit by building practices on them. For example, from the belief about involvement comes the practice of involving everyone in the workforce regularly in "doing things in safety" (such as working on a rules and procedures committee or taking responsibility for the next safety meeting). The key practices are found in all of the very safe companies, although the emphasis varies.

An initial list of beliefs and practices was derived from observation of the very safe companies and from the author's management and consulting experience. As the research proceeded, the list was extended and improved.

Together with the framework, the beliefs and practices form a model that can be used as the basis for measurement of the state of safety management. The model is also the road map for the agenda for safety excellence. Thus it is essential that the set of beliefs and practices be thorough, precise, and comprehensive.

The list of 25 beliefs and practices has been organized as follows:

1) The 10 fundamental beliefs
   a) The 5 fundamental general beliefs
   b) The 5 fundamental beliefs about safety management
2) The 15 specific safety practices and the beliefs that underlie them

2.3.1 The Beliefs and Practices for Excellence in Safety

2.3.1.1 The Five Fundamental General Beliefs

1. The health and safety of people has first priority and must take precedence over the attainment of business objectives.
2. All injuries and occupational illnesses can be prevented. Safety can be managed and self-managed.
3. Excellence in safety is compatible with excellence in other business parameters such as quality, productivity, and profitability; they are mutually supportive. Safe, healthy employees have a positive impact on all operations. They have a positive effect on customers and enhance credibility in the marketplace and in the community.
4. Like quality, safety must be made an integral part of every job. "Do it right the first time."
5. Good safety is "mainly in the head." Most injuries and safety incidents occur because of inattention, not because of lack of knowledge or for physical reasons. People take risks because they believe that *they* will not be hurt.

2.3.1.2 The Five Fundamental Beliefs About Safety Management

6. Top management must be committed to excellence and drive the agenda by establishing a vision, values, and goals; by seeing that all line managers have safety improvement objectives; by auditing performance; and by visible personal involvement.
7. Safety is a line responsibility. Each executive, manager, or supervisor is responsible for and accountable for preventing all injuries in his or her jurisdiction, and each individual for his or her own safety and, in a less direct sense, for the safety of co-workers.
8. Involvement of everyone in "doing things in safety" is the most powerful way to embed safety values and to build safety awareness.
9. Safety training is an essential element in developing excellence.

It complements but cannot replace "learning by doing" (in itself a method of training).

10. An organization committed to safety excellence will have a broad array of safety systems and practices, thoroughly and conscientiously implemented with broad workforce participation.

2.3.1.3 The Fifteen Specific Safety Practices and the Beliefs That Underlie Them

11. **Safety Meetings.** Regular, effective safety meetings involving all people in the workplace are an essential part of good safety.

12. **Safety Rules.** Comprehensive, up-to-date safety rules, crafted with broad participation and consistently applied, are essential for excellence in safety and also assist in doing the job well.

13. **Enforcement of Safety Rules.** Disciplinary action for safety infractions is an essential part of good safety. Its intent is not punishment or retribution but the correction of unsafe behaviour, the demonstration of the standards of the organization, and the weeding out of those who will not accept their responsibility for safety.

14. **Injury and Incident Investigation.** Every injury or incident is an opportunity to learn and improve. Thorough, participative processes of investigation are a cornerstone of safety.

15. **Workplace Audits-Inspections.** Auditing the workplace for physical conditions, for the effectiveness of safety systems, and for the safety awareness of the people who work there is a valuable way to improve safety.

16. **Modified Duty and Return-to-Work Systems.** Excellence in safety is enhanced by thorough efforts to find modified duties for injured people who cannot do their regular jobs but who can safely do other work and by comprehensive initiatives to assist in rehabilitation and ensure early return to work.

17. **Off-the-Job Safety.** The organization has a responsibility to promote "off-the-job" safety as well as safety in the workplace, to help make safety "a way of life."

18. **Recognition for Safety Performance.** Recognition for safety achievement and celebration of safety milestones provide strong reinforcement for the organization's commitment to excellence.

19. **Safety of Facilities and Equipment.** The containment of hazards by integrating leading edge safety technology into the design and operation of facilities is essential for outstanding safety.

20. **Measuring and Benchmarking Safety Performance.** Comprehensive, up-to-date safety statistics, communicated to all, are a cornerstone of safety management. Benchmarking against the best will help improve safety.

21. **Hiring for Safety Attitude.** Safety can be enhanced by hiring people with good safety values and attitudes.

22. **Safety of Contractors and Subsidiaries.** Contractors and subsidiaries, including foreign operations, must work to the same safety standards as the company.

23. **Involvement in Community and Customer Safety.** Excellence in safety will lead naturally to involvement in and leadership in community and customer safety. This will be valuable to the organization and to the community and customers.

24. **The Safety Organization.** The safety organization is a valuable asset in attaining excellence in safety. It should be led by the line organization with broad participation by the entire workforce, particularly those at the working level.

25. **Safety Specialists.** Safety specialists can provide valuable assistance to the safety organization. They must avoid taking responsibility for managing safety or accepting accountability for results; these lie with the line organization. Rather than doing the work themselves, they should facilitate involvement of the workforce.

The beliefs and practices are a critical part of the model, and they form the main base from which the safety questionnaire was developed. Thus they require in-depth explanation. To avoid repetition, this is done in Chapter 6, where the belief behind each question is discussed before the research results from the use of that question are presented. Thus only a brief introduction to the beliefs is given here.

2.3.2 The Five Fundamental General Beliefs

The ten *fundamental beliefs* form the foundation for excellence in safety.

The first five beliefs set the safety culture of the organization. They are the essential indicators of management commitment. None leads to specific practices; instead, they influence *all* practices. If these beliefs are endorsed by leaders, strongly held throughout the organization, and brought to life in the practices, safety will be at a high level.

The first belief, that safety must have overriding priority, influences all aspects of safety. Although it appears simple, it has profound implications. The priority given to safety must be visible in all the actions of the company, particularly in the behaviour of management. This is a critical measure of management commitment.

The belief that all injuries can be prevented also has broad implications. If all injuries *can* be prevented, it becomes incumbent on leaders to see that they are. This also means that injuries can never be blamed simply on worker negligence, thus supporting the more specific belief about line responsibility.

The belief that safety excellence has a positive effect on business performance is critical. Without it, cost-benefit tradeoff thinking leads to compromises in safety.

The belief that safety must be made an integral part of every job is a parallel to the belief that has helped drive quality to new levels.

The last of the general beliefs, that "good safety is mainly in the head," leads to important insights. Unless leaders understand that excellence starts with the thinking of people, particularly with their own thinking, they will continue to rely on improvements in physical things and safety systems and will continue to delegate safety to subordinates.

2.3.3   The Five Fundamental Beliefs About Safety Management

The second five beliefs deal more specifically with safety management, and some specific practices flow directly from them. Yet they too are fundamental: they underlie the safety culture and affect all practices. They lay out the agenda for top management, for line management, and for the workforce in managing safety.

The first of these beliefs defines the responsibility of the CEO and the top leaders to see that a vision of excellence is created and that specific safety values are developed and thoroughly understood throughout the organization. The leader must see that goals and objectives are established and must audit performance against them. The leader must be visibly involved in safety.

The second belief defines the responsibility of line managers for the safety of their people and their accountability for injuries that occur to them. Line management plays a central role in driving the safety agenda, translating the vision, values, goals, and objectives into practice.

The third belief establishes the importance of involvement as a vehicle for developing commitment to safety values and promoting safety awareness and understanding of safety practices. It calls for much deeper involvement of everyone in "doing things in safety" than is the case in most companies.

The fourth belief about safety management defines the necessity for thorough training of the workforce in safety while stressing that training is not a substitute for involvement but complementary to it.

The last of the beliefs about safety management defines the need for a comprehensive set of safety systems and practices, regularly updated and meticulously implemented, and for broad participation of the entire workforce in managing the practices. It thus leads directly to the important safety practices.

2.3.4   The Specific Safety Practices

The fundamental belief that underpins each of the 15 most important practices is included here to help reveal the underlying purpose of the practice. The belief is more fundamental than the practice and may have broader implications. For example, the belief in the power of recognition for safety achievements in reinforcing excellence leads to specific recognition practices, but it must also be a day-to-day responsibility of line management. It must also be understood by workers and used by them in the safety activities in which they participate, for example, in the safety promotion activities of an off-the-job safety committee.

The practices can be expanded further into a more comprehensive set of specific safety practices. For example, the practice relating to "Measuring and Benchmarking Safety Performance" calls for "a comprehensive array of measurements, kept up-to-date, compared to objectives, benchmarked against the best performers, and communicated to all of the workforce." In the very safe companies, this practice will be supported by a set of specific practices such as:

- Measurement of leading and coincident as well as lagging indicators
- Systems to assess attitudes to safety (e.g., perception surveys)
- Systems and practices to communicate information quickly to all

2.4   THE COST-BENEFIT TRADE-OFF

One of the most important barriers to excellence in safety is the belief of many leaders that, beyond achieving a certain level of safety, further efforts will cost more than they deliver in economic benefit. This cost-benefit thinking emerges so often in the assessment of safety management that it warrants a separate discussion.

The safety literature often refers to the cost-benefit trade-off in safety. The theory says that as effort and money are invested to improve safety, the point of diminishing returns is reached, beyond which the effort is better spent elsewhere, in improving costs, productivity, or quality. Bequele (18) stated the concept as follows:

> In adopting safety and preventative measures, however, the firm has to balance their costs against their benefits. . . . The firm will continue to invest in safety or prevention until it reaches the point where its cost is equal to the additional saving from a reduction in accidents. Formally speaking, that happens to be the point which will also minimize the overall costs, i.e., where the marginal cost of safety is equal to its marginal benefit.

The cost-benefit balance is sometimes illustrated graphically, as for example in a British Health and Safety Executive Report (19).

In Fig. 2-2, the apparent minimum cost is simply the sum of the cost

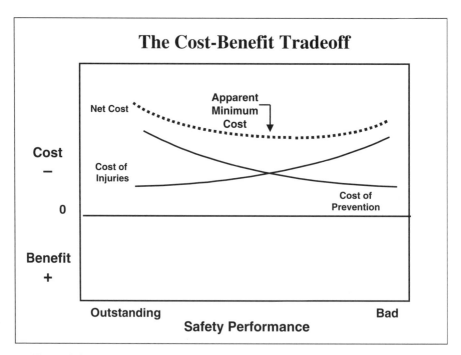

Figure 2-2 The cost-benefit tradeoff, as schematically shown in the literature.

of injuries and the cost of prevention. These schematic representations always seem to show the minimum in the middle. The author has not seen a graph with real data, which would be very difficult to get.

Presenting the theory in this way implies that the best compromise is in the middle, at average safety. It is not surprising that many leaders believe, or behave as though they believe, that the appropriate trade-off is at what could be called *mediocre* safety. Figure 2-2 portrays such thinking well.

The cost-benefit theory is an obvious truism. If safety performance is truly bad, the cost of injuries is very high and provides an important incentive to improve. However, beyond attaining average performance, cost-benefit thinking would cause leaders to question why a company should strive to do better. Attaining very good or outstanding safety performance would likely cost more than it benefits the organization.

This assumption is not contradicted by the literature; there is little direct evidence that excellent safety can be combined with outstanding business results. By concentrating on the measurable direct and indirect costs and benefits, researchers have inadvertently helped to convince corporate leaders that they should go no further than the mediocre.

The analogy between creating excellence in safety and developing world class quality is striking. It has been discussed in the literature (see for example Refs. 20 and 21), yet it has not registered fully with corporate managers. Many of the concepts are common to both: the need to establish commitment at the top, the need to instill the principles in the mind of everyone, the requirement that concrete objectives and indicators of success be established and measured, and the meticulous attention to excellence in safety practices.

Quality, safety, and integrity are not activities in themselves; in the best performing companies, they are fundamental values, the way that everything is done and done right the first time. As corporations move steadily towards self-management, it becomes even more important that safety, like quality, be managed in a different way.

If the only reason for seeking excellence in quality were avoidance of the obvious costs of defects, the great improvements in quality achieved in the last couple of decades would not have occurred. A generation ago, the conventional practice was to provide the customer with quality no better than the stated need—to meet the customer's specification. Doing better would cost the supplier and would not benefit the customer. This is exactly the same theory—the cost-benefit trade-off in quality. Japanese companies were the first to show how wrong that premise was, not by theory but by the demonstration that world-leading quality does not cost but, instead, provides enormous benefits. Worldwide, corporations followed this lead. CEOs have come to realize that without excellence in quality, the competitive position of their companies would be weak whether or not they could ascribe specific immediate benefits to the quality drive.

Extend these concepts to safety. Although there are direct and indirect costs and benefits to improving safety, there are also very much larger *intangible* costs and benefits. Can the dedication, initiative, and meticulous attention required for exceptional quality be established in a workforce in which people are routinely injured? Aren't the orderly thinking, the concentration on doing it right the first time, just what is needed to avoid defects in safety (injuries)? Viewed thus, the cost-benefit graph takes on a different form. In Fig. 2-3, a line was added to represent the intangible costs and benefits of safety. The net cost line is radically altered, and the true minimum is moved far to the left. Where is the real minimum cost of safety? There is undoubtedly an upturn, at the extreme left end of the graph, as perfect safety is

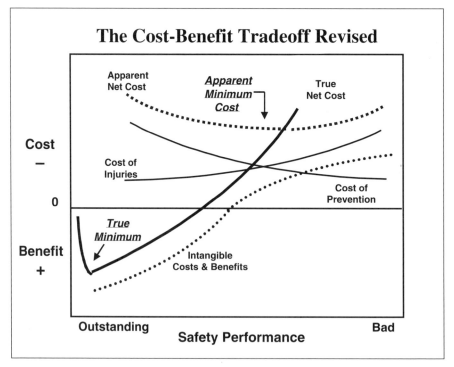

Figure 2-3   The cost-benefit tradeoff revised to consider the intangible benefits of excellent safety—in quality, productivity, teaming, etc.

approached. The issue is the shape of the curve and the point at which this minimum is reached.

Companies that have achieved outstanding quality have worked from the premise that there is no *practical* limit to how well they can do. They strive for six-sigma levels (single figures in defects per million) or even for zero defects. Some are achieving these previously unbelievable levels of quality, and their costs continue to decrease. Similarly, the world's safest companies believe that "all injuries and occupational illnesses can be prevented." They hold that there is no practical limit to the level of safety that can be achieved without undue costs and in the long run will result in net benefits. Safety is built in, values driven, part of the culture, integrated into everyone's job all the time. To paraphrase the quality credo, "safety is free."

One of the most powerful refutations of conventional cost-benefit thinking can be found in the dramatic spread in safety performance among companies. The data used in this research is plotted in Fig. 2-4. Most of the data are average data for 5 years.

The horizontal scale of the graph is logarithmic: each quadrant represents performance 10 times better or worse than the adjacent quadrant. The measure is 5-year average lost work injury frequency (LWIF). The safest companies have records 1000 times better than the worst and more than 100 times better than the average! Had DuPont Canada performed at the average of Ontario manufacturing in the five years from 1993 through 1997, it would have had more than 500 lost work injuries in the period rather than 5. From 1971 through 1991, the Kanagawa plant of Hitachi Japan went almost 20 years with 2500 employees without a lost work injury. It set the record for Japanese plants—an astounding 116 million hours without a lost work injury.

The five very safe companies in this survey have come close to eliminating major injuries. And they are no slouches in profitability: they are leaders in their industries.

DuPont Canada's safety-profitability record, updated from the previously mentioned paper (3) is shown in Fig. 2-5. The data from one situation do not, of course, prove the case. However, they do demonstrate the possibility of achieving excellence in safety as profitability is improved and sustained at high levels.

Attempts to correlate safety performance with input factors have not been very successful. The author compared operating profit with safety for the companies in the Canadian Chemical Producers' Association (CCPA) for the five years from 1986 through 1990. Sufficient data was available for 32 companies, for which the safety performance in total recordable injury frequency (TRIF) ranged from 0.8 to 30. There was

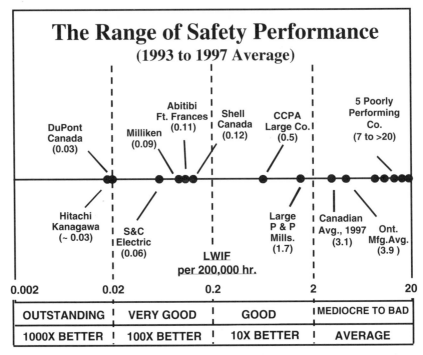

Figure 2-4   The surprising 1000 fold range in the safety of companies.[vi]

a great deal of scatter in the data, and, although better profitability did go with better safety, the correlation was statistically weak.[vii]

In a recent paper, the data in the literature on this subject were reviewed (22), with the conclusion that most of the information was qualitative. The results of a new study of the relationship between profitability and safety were also reported. The study indicated that of the six classes of company assessed, in four classes the companies with poor safety had somewhat better average profitability than those with good safety and in two classes, profitability was better for the safer companies. The connection appeared weak, however: 9% profit as a percentage of sales was considered "better" than 8.3%.

A quantitative relationship between quality and profitability has been demonstrated only recently. One of the most powerful proofs comes from the work of the Strategic Planning Institute (the PIMS database; Ref. 11). Over many years they have collected data on several

---

[vi] Data for the 5 poorly performing companies is for various 1- to 3-year periods from 1993 to 1997. The Canadian average is for 1997. The Hitachi data is approximate for the mid-1990s. The rest is 1993–97 average.
[vii] Unpublished analysis derived from CCPA data.

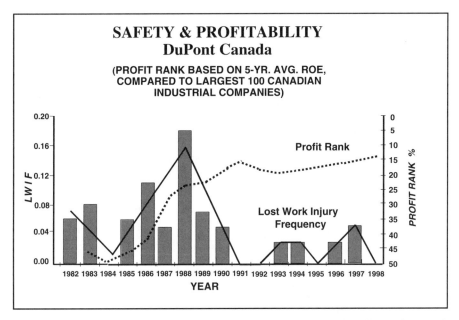

Figure 2-5 DuPont Canada: outstanding safety in the 1980s, virtually no lost work injuries in the 1990s.[viii]

thousand businesses. With such data, they have been able to isolate statistically the effect of one variable on results. Their research shows the strong positive effect of customer-perceived quality on profitability. Similar long-term, multibusiness information would likely show a strong positive effect of excellence in safety on profitability. This would be a difficult task, but it would refute the simple cost-benefit thinking that hampers safety improvement today. Until this is done, we have only indirect and case data to support the contention that "safety is free."

The key to major improvement in safety is to develop understanding, commitment, and will in corporate leaders, CEOs in particular. They must be convinced that excellent safety will not cost more in the long run and will deliver valuable results. Without such understanding, the commitment and will to act are missing. Today, many leaders think that "juggling all the balls" is neither feasible nor profitable. For excellence in safety, leaders must have the passion for excellence that has been the cornerstone of outstanding quality.

[viii] The graph corresponds to the period the research was done. DuPont Canada has since extended its excellent record, with no lost work injuries in 1999 or 2000.

2.5 LIMITATIONS OF THE MODEL

The model has been derived from the management methods of companies that have excellent safety records and has been discussed extensively with them. It is not too likely that a critical element is missing. There are, however, two basic issues that the model does not address directly. First, safety management is an integral part of the overall management process and is influenced by the environment of the organization. Companies have different levels of hazards, different types of staffing, and different organizational cultures. The model deals with safety management *only*, not with these other important organizational influences. Second, the model does not tell much about the relative importance of the practices.

2.5.1 The Organizational Environment—Hazards, Staffing, and Culture

It may be more difficult to achieve safety excellence in operations that are inherently hazardous, but the injury record should not depend primarily on the extent of the hazard. The excellent record of some companies that operate in hazardous industries and the poor safety record of many that have mainly office environments are evidence for this view.

The type of staffing (temporary or permanent, skilled or unskilled) depends very much on the industry, and this is sometimes blamed for poor safety. The construction industry is often cited as combining the most difficult staffing characteristics with one of the most hazardous environments. Many construction workers are temporary, at least to the job or the project if not to the construction organization. Some of the jobs are unskilled, and some have significant physical hazards. Yet there is no doubt that large construction jobs have been completed with very low injury rates. In the late 1980s, DuPont Canada built a $100 million chemical plant in western Canada on a new site. At times there were many contractors with several hundred tradespeople on the site, and work continued for 2-1/2 years through the northern winters. In over a million hours of exposure, there were no lost work injuries. In a Worksafe Australia case study, the safety performance in the construction of a multibillion dollar natural gas project was reviewed (23). In the first phase of the project, there were several hundred lost work injuries and two deaths. In the last phase, after a major improvement in safety management (involving Shell International), there were no lost work injuries in 3 million hours of exposure over 2-1/2 years. In both the

DuPont and the Shell cases, excellent safety was developed using the concepts of the model of safety management described here. Some industrial companies deal with fairly large numbers of temporary workers, such as students for summer vacation relief, without increased injury rates. It is the author's opinion that special staffing situations, like the extent of the hazard, present special difficulties but do not prevent the development of excellence in safety.

The effect of the work environment and organizational culture is an entirely different thing. Can there be excellent safety when there is poor morale and/or worker-management conflict? Can an excellent work environment exist when there is little attention to safety and a bad injury record? These factors were not taken into account in developing this model. To what extent is excellent safety or very poor safety due to the work environment and to what extent to the factors in the model?

Envisage two future scenarios for an organization with very poor safety and a poor workplace environment (perhaps low morale and/or poor labour-management relations).

In the first scenario, management does not attempt to improve the workplace environment but undertakes a determined, long-term drive to improve safety, installing all the concepts and best practices of the model. Undoubtedly safety will improve, possibly even to very good levels. However, it would be difficult to achieve excellence unless the workplace environment improves, although improvement will occur simply as a by-product of management working to sounder principles of safety and because of the much greater workforce training and involvement. (Working on safety is perhaps the best place to begin a process of involvement, even when the goal is more general, perhaps a transformation from a hierarchical organization, to a more self-managing organization.)

In the second scenario, management undertakes to improve the workplace environment by developing more open communication, involvement, and empowerment but does not address safety improvement directly. It is difficult to conceive of a major improvement in the workplace environment being possible while people are being injured frequently. Thus, in practice, safety will have to be brought at least to the average level as part of improving the workplace environment.

These scenarios illustrate how intertwined safety is with the workplace environment. It is clear that the workplace environment *alone* does not cause excellence in safety. That requires working directly and purposefully on the concepts and practices used by the very safe companies, those described in the model of safety management.

2.5.2 The Relative Importance of the Practices

The model is constructed so that the five fundamental factors are all interrelated and necessary (management commitment, line ownership, involvement, comprehensive safety systems, and the safety organization). Some practices are clearly more important than are others, and the model does not tell us much about their relative importance (nor was this attempted in the research). Here we must fall back on experience and judgement.

The practices involving safety rules and injury investigations are examples of those that are probably essential for excellence in safety. Some others are less essential. For example, very good workplace safety can probably be achieved without a focus on off-the-job safety, but, not surprisingly, the very safe companies have found benefit in having strong off-the-job programs. By and large, the package of practices hangs together as the necessary combination to help a company advance to world class excellence.

3

THE SAFETY QUESTIONNAIRE

3.1 BACKGROUND TO THE DEVELOPMENT
OF THE QUESTIONNAIRE

The factors in the model of safety management, such as management commitment, cannot be directly observed or measured. Although the beliefs that underpin the factors cannot be directly measured either, they are narrower in scope and more specific than the factors. The extent to which the beliefs are held can be perceived in a general way by observation in the workplace and by interviews with executives and workers. The presence and quality of safety practices can also be assessed by observation, but it is often difficult to determine how effective these practices are without observation over a period of time. Even then, observation may not give a true picture.

Observation and interview techniques suffer from serious handicaps. They are subjective both in the judgements made by the observer and in the impressions given by those observed or interviewed. It is difficult to compare observations of one organization to those of another or to compare observations made in the same organization at different times. How can the effectiveness of injury investigation in two very different companies be compared by observation? By focusing on the views of a few people, observation and interviewing techniques can yield a distorted picture. It is difficult to avoid overweighting the views of management and supervision. A better way to determine the presence and effectiveness of the key factors, beliefs, and practices is needed.

Perception surveys have been used for many years in various applications. They have been used, but less frequently, in safety management studies (see, e.g., Refs. 12–14). The author is not aware of a comprehensive questionnaire based on a structured model of safety management or of quantitative benchmark data obtained by surveying a cross section of people in companies with excellent safety and companies with very poor safety.

In the early 1980s, the DuPont Company used a safety survey internally to assess the state of safety management in some of its US plants. The author remembered that this had been done, but no information seemed to have remained on it in the DuPont files.

The industrial psychologist Walter R. Mahler pioneered the application of perception surveys to managerial behaviour, using them as a core tool in his Advanced Management Skills Program. In this executive course, the perceptions held by subordinates of their managers' management style and behaviour were collected for a comprehensive range of factors. Over many years, a large database was developed against which individual results could be assessed (24).

In earlier consulting work, the author had developed a questionnaire to assist in the diagnosis of safety issues at a pulp and paper mill (25). It was based on the model of safety management presented here, particularly on the beliefs and practices. In building the questionnaire, the author was influenced by Mahler's techniques, particularly the idea of developing a database of answers to standard questions—in a sense thus calibrating the questionnaire. The questionnaire was improved in subsequent consulting projects. It became the main tool in this research project. Although some changes were made during the research, it has remained close to the form developed for consulting.

Well-designed questionnaires completed by a cross section of an organization can provide better, more quantitative information on a more comprehensive range of subjects than can be gained by observation or by interviews. They reveal how strongly beliefs are held and the extent to which they influence behaviour. Questionnaires deliver a quantitative report card on the practices. They assess perceptions, exposing gaps between what is visible, what is said, and what people think really happens. Questionnaire results can also provide a solid base for planning later observations, interviews, and focus groups. The combination of questionnaires with observations and interviews reveals much more than either technique alone.

The concepts of customer-perceived quality were used in designing the questionnaire. The intent was to be *objective, comparative, quanti-*

tative, and *confidential*. For example, the priority a company gives to safety is critical. Leaders and safety specialists in the organization cannot be objective on this issue. Many will say that they give safety high priority. They may indeed think they do. Much more objective information can be gained by asking people in a confidential questionnaire to rank the relative priority that they think management, supervision, and workers give to safety, production volume, quality, and costs. Managers may answer that they give priority to safety. If they really do, workers will agree. If instead, the company gives precedence to costs and productivity, the discrepancy between the stated priorities and the real ones will be exposed. The perception of priorities held by the workforce will reveal what really governs safety action.

Contrast this with the information gained by observation and interviews of managers and workers. In a company where safety is poor, managers will probably still say that safety is given at least fair priority. Others may not support this statement but may be reluctant to say so openly. The researcher or consultant is not in a good position to dispute this, even if there are strong indications that safety has low priority. In a confidential questionnaire, however, workers and many staff members will likely say that safety is near the bottom of the priority list. Interviews and focus groups can then be used to confirm whether this is indeed the case and to seek out the underlying reasons for it.

A company may claim to have good safety rules, and management may believe they are followed. This is a key practice, based on important beliefs. Yet it is difficult to determine by observation whether rules are followed, although questioning workers verbally may give an indication. A far better assessment can be made by addressing a carefully designed question anonymously to a cross section of the workforce. To be as nonleading as possible, the question should not be "Are safety rules obeyed?" but something like this:

To what extent are the **safety rules** of your organization **obeyed**? Check only one answer.

$\square_1$ All safety rules are obeyed without exception.

$\square_2$ People generally obey the safety rules.

$\square_3$ The safety rules are guidelines, sometimes followed, sometimes not.

$\square_4$ The safety rules are often not obeyed.

$\square_5$ People pay little attention to the safety rules.

The choice of five responses makes the results much more specific and facilitates comparison. It is one thing to report to a plant manager that you have observed that safety rules are not obeyed or that people say that they are not. It is far more persuasive to report that 75% of the 40 workers polled answered that "the safety rules are often not obeyed," especially if this finding then leads to focus groups and interviews that reveal the background of the issue and expose examples.

For the questionnaire results to be statistically valid, a sufficient number of people must be surveyed. For example, in a workplace with 300 people, the survey might include about 70 people—the leader and the senior managers (usually 6–12 people), 10 of the 20 supervisory staff, and 50 workers. It is particularly important that the questionnaire poll more than the leadership. The leaders may feel that they hold certain beliefs, and they may say so in very definite words. The rest of the organization may have a different view: they may not recognize the beliefs or see them translated into action.

It is critical that the respondents are assured of anonymity. Otherwise, workers in particular will not participate.

Four levels of insight can be gained from safety questionnaires:

1. The absolute level of the answers yields important information. For example, if most of the answers indicate that safety rules are often not obeyed, there are serious safety problems.
2. Comparison among the answers from management, supervision, and workers yields important insights.
3. The results from the subject company can be compared with those of other companies, particularly those with excellent safety and those with poor safety.
4. The results obtained at one time can be compared with those obtained later to help determine whether improvement has occurred.

In summary, the information garnered by structured questionnaires can be much superior to that gained by observation or interviews, regardless of how carefully they are done or how skilled the auditor. By identifying the true situation, the results can lead to more focused and useful observation and verbal questioning.

3.2 SCOPE OF THE QUESTIONNAIRE

The questionnaire used here was designed to address all of the important factors, beliefs, and practices in the model. Over several years of

The Relationship Among Beliefs, Factors and Questions

	BELIEF	QUESTION	FACTOR*
	THE FUNDAMENTAL BELIEFS		
	The Fundamental General Beliefs		
1	*The Priority Given to Safety -- priority individuals give to safety*	1	MC-ALL
	-- priority people say others give to safety	2	MC-ALL
2	*The Belief That All Injuries Can Be Prevented*	3	MC-ALL
3	*The Interaction Between Business & Safety -- effect of safety drive*	4	MC-ALL
	-- cost benefit break-point	5	MC-ALL
4	*The Extent That Safety Is Built In*	6	MC-ALL
5	*The Belief That Good Safety Is Mainly "In the Head"*	---	MC-ALL
	The Fundamental Beliefs -- Safety Management		
6	*Management Commitment, Vision, Values, Goals & Objectives*	7	MC-ALL
7	*Line Management Responsibility & Accountability for Safety*	8	LO-ALL
8	*Involvement In Safety Activities -- perceived involvement*	9	I&T-ALL
	-- perceived empowerment	10	I&T-ALL
9	*Safety Training*	11	I&T-ALL
10	*The Scope and Quality of Safety Practices*	---	SS&P-ALL
	SPECIFIC SAFETY PRACTICES & THE BELIEFS THAT UNDERLIE THEM		
11	*Safety Meetings*	12	SS&P
12	*Safety Rules -- quality*	13	SS&P
	-- extent rules are observed	13	
13	*Enforcement of Safety Rules*	14	SS&P
14	*Injury & Incident Investigation*	15	SS&P
15	*Workplace Audits-Inspections*	16	SS&P
16	*Modified Duty & Return-to-Work Systems*	17	SS&P
17	*Off-The-Job Safety*	18	SS&P
18	*Recognition for Safety Performance*	19	SS&P
19	*Employing the Best Safety Technology*	20	SS&P
20	*Measuring and Benchmarking Safety Performance*	21	SS&P
21	*Considering Safety Attitude in Hiring*	---	SS&P
22	*The Safety of Contractors & Subsidiaries*	---	SS&P
23	*Involvement in Community & Customer Safety*	---	SS&P
24	*The Safety Organization*	22	SO&SP
25	*Safety Specialists*	23	SO&SP
	Satisfaction with the Safety Performance of the Organization	24	---
	* Key to Factor Designation (see figure 2-1)		
	MC Management Vision, Commitment & Drive		
	LO Line Ownership of Safety		
	I&T Involvement in Safety Activities, Training		
	SS&P Safety Systems & Practices		
	SO&SP Safety Organization, Specialists		
	ALL Pertains to All Above Factors		

Figure 3-1   The relationship among the beliefs, the factors in the model and the questions in the questionnaire.

consulting and through this research, it has been refined, some questions improved and others deleted. The final questionnaire has 24 questions—enough to cover the important safety concepts but short enough that respondents will complete it. It is included in Appendix E.

The relationship among the factors in the model, the beliefs and practices, and the questions in the safety questionnaire is illustrated in Fig. 3-1.

Although the questionnaire covers almost all of the important issues in safety management, there are some that cannot be assessed well with this technique. The belief that "good safety is mainly in the head" leads to important insights about safety. However, it does not lend itself to a direct question. Including it as a belief leads to discussion and understanding, and the belief is inherent in many of the other questions.

Much of the value of the questionnaire comes from the perspective gained by having a cross section of the organization, particularly workers, answer the questions. However, a few of the practices would not be well-enough known by many respondents for valid answers to be obtained. For example, workers may not be familiar with the safety considerations in company hiring practices, with the safety of subsidiaries, or with the involvement of employees outside the company in customer and community safety (beliefs-practices 21, 22 and 23). These topics can be handled better by interviews and focus groups.

Even the best-designed questionnaires cannot give complete answers to complex issues. There is a balance between having the questionnaire short enough to keep the attention of respondents and at the same time comprehensive enough to yield complete answers. For example, management commitment to safety is critical. If workers agree that their leaders give high priority to safety, if there are well-recognized safety values, and if line management is perceived by workers to take full responsibility for safety, there is certainly strong management commitment. All of these elements are included in the questionnaire. However, it is also useful to explore the clarity of the safety goals, the thoroughness with which objectives are established by each manager and supervisor, the extent to which achieving objectives is reflected in managers' pay, the extent of direct involvement of the leaders, and other such items. These elements can be assessed by investigation at the workplace, by interviews, and by discussion in focus groups.

Thus, although the safety questionnaire was the most important tool in the safety survey conducted in the five very safe companies, it was supplemented with an organized framework of investigation and observation and with interviews with key people and focus groups.

The procedures for administering the questionnaire are more fully described in Chapter 5, Research Methodology. In the presentation of the results in Chapter 6, the relationship between the beliefs and the wording of the questions is elaborated further.

4

SELECTION OF COMPANIES FOR RESEARCH

The original intent of the research was to investigate the safety management practices of 7 to 10 very safe companies in depth, rather than to survey a larger number of companies with a range of safety records. They were to be mainly Canadian companies but if possible should also include examples of US, Asian, and European firms. Although the focus was mainly on companies in the goods-producing industries,[i] the intent was to cover a range of business sectors, including an example of a company with excellent safety in a largely office environment. The research was intended to cover some of the safest companies in the world, but to have examples of important Canadian industry sectors, some of the companies included might not meet that standard. However, they would have "excellent" safety records and be leaders in safety in their sector in Canada.

The first hurdle for qualification as having "world class" safety was a 5-year average lost work injury frequency (LWIF) no worse than 0.25 per 200,000 hours, with no single year worse than 0.50. This is 15 times better than the Ontario manufacturing average for 1993–1997. (The five companies chosen had average LWIFs from 0.03 to 0.12 or 30–130 times better than the Ontario average.) The company also was to have

---

[i] Including manufacturing and natural resource extraction and processing (e.g., mining, pulp and paper, oil and gas, and food processing). The goods-producing sector has substantial safety hazards. It is also a sector from which many of the examples of outstanding safety have emerged.

an excellent record in total injury frequency and no fatalities in the 5-year period.

The use of LWIF as an identifier of excellence is disputed by some. It is a simple indicator but a good one. A low LWIF record does not necessarily indicate excellence, but there cannot be excellence without it. In larger companies, it is a critical distinguishing factor, the safest companies having LWIF records up to 1000 times better than the worst. The very safe companies do not use LWIF as their main *control* indicator because they have largely eliminated such major injuries. Instead they use a variety of indicators such as total injury frequency (TRIF), first aid case frequency, and the frequency of incidents and near misses. To these lagging indicators, they are increasingly adding coincident and leading indicators such as perception surveys, safety activity indices, quantified audits, and other such measures.

After preliminary screening on a LWIF basis, a variety of other safety performance data for the company was assessed before research was initiated. TRIF was an important indicator. The safest companies—those with truly world class safety records—have long-term average TRIFs consistently below 1.0. Very few companies meet this standard. Three of the five very safe companies in this research had TRIFs below 1, and the other two were at about 2.

The company's safety culture and programs were reviewed to ensure that the safety record was achieved by purposeful action and not just as a result of a situation of low risk or failure to record injuries. The intent was to identify companies that ranked consistently among the best in long-term safety performance and that also had good business performance records.

In this book, "safety" is used as a short form for safety and occupational health. It is a valid criticism that occupational illnesses were not used specifically as a determinant of excellence. Lost time or restricted time caused by occupational illness is included in the US Occupational Health and Safety Administration (OSHA) standard for statistical reporting. The effect of occupational illness shows up in the record, however, only when the illness becomes serious enough that time is lost (not all the companies in this research follow OSHA guidelines). However, it is the author's observation that companies that are excellent in safety are also meticulous in striving to anticipate and prevent occupational diseases.

The safest companies espouse the belief that "all injuries can be prevented." They do not accept that the particular hazards of their sector or of any specific operation excuse injuries or allow a lower standard. The chemical and oil company sectors, for instance, have much better than average safety records despite the hazards of their operations. Some companies with mainly office operations have poor safety records. Nevertheless, the safety performance of the particular company compared with its sector peers is an important criterion. The comparative safety record is particularly important in the case of off-shore companies, where the measurement criteria may be different from that in North America. It is also important for sectors in which many companies have good safety records and in sectors that should be covered because of their importance to Canada but in which no companies have truly outstanding safety (e.g., pulp and paper).

As a study of leading edge safety management, the first criterion was consistent excellence in safety. The original intent was also to establish criteria for business excellence, albeit less stringent and more subjective criteria such as above-average long-term records in profitability and quality. It soon became apparent that the main difficulty in the research was to identify companies with world class safety and then to persuade them to participate in the study. Not unexpectedly, the very safe companies chosen for study all had good business performance records. Thus no specific business performance criteria were established.

4.1 SIZE OF COMPANY

A LWIF of 0.25 represents the following number of injuries for various sizes of companies:

No. of People	Injuries per yr	Injuries in 3 yr	Injuries in 5 yr
10,000	25	75	125
1,000	2.5	7.5	12.5
200	—	1.5	2.5
100	—	0.75	1.25

It is difficult to identify true safety excellence in small companies from their LWIF record. It takes a long time to confirm whether the apparently good record of a small company is a statistical anomaly or whether its safety performance is truly a standout. A company with 100 employees would have to have no more than one injury in 5 years to be considered. Some smaller companies and units of companies do this. For example, the DuPont Canada Research Centre, with about 150 people, has not had a lost time injury in over 30 years. There are similar problems in ascertaining whether the performance of small companies is poor. As a practical limit, 100 people was the minimum company size considered for this research. The smallest of the 10 companies for which detailed results are reported had about 250 people.

The inability to be sure of the quality of safety in smaller companies (from statistical records) does not mean that the findings of this study do not apply to them. The model of safety management flows from fundamental principles of human behaviour that are not dependent on the size of the organization.

The intent of this study was to identify excellent *companies*, not excellent units in companies. If the values and practices are deeply entrenched, the company will see that excellent safety exists in all of its operations. This criterion was "bent" in the case of the pulp and paper sector. At the time of this research, none of the Canadian pulp and paper companies had world class safety records but some *mills* did have very good records. Because of the importance of the sector in Canada, one of these mills was included in the study.

4.2   DIFFICULTIES IN IDENTIFYING COMPANIES WITH EXCELLENT SAFETY

A great deal of difficulty was experienced in identifying companies with world class safety performance records.

1. By definition, there are few such companies anywhere and thus very few in Canada.
2. There is no comprehensive national database of corporate safety data in Canada. The provincial data refer to sites, not companies.
3. Although there are some national data for the US, they are not comprehensive.
4. It is very difficult to identify offshore companies without spending significant time in those countries.
5. Individual companies with good but not outstanding safety records usually are unaware of how their safety compares with that of the truly safe companies. Thus many false leads were suggested.

Because the research was partially supported by Canadian governments, most of the Canadian companies that were approached agreed to participate, although a few refused. It was a different story outside of Canada, where companies unfamiliar with the researcher were reluctant to commit to an in-depth survey of their safety.

4.3 CANADIAN COMPANIES

The identification of Canadian companies with world class safety was frustrated by the lack of a national record system. The provincial agencies record the safety of sites, not companies. Their orientation has been toward the poor performers, not the good ones. The provincial Workers' Compensation Boards (WCBs) and like agencies, all sponsors of this research, tried to help. The companies that they identified were often local sites of international companies, already known to the author, or medium-sized Canadian companies in the services sector, such as software shops. A dozen of the latter were assessed and found not to have identifiably strong safety performance. With small, largely professional staff involved in clean office work, safety was not a focus for them. It was suspected that if an injury occurred, the individual would usually just go on with the work, or take time off without formally recording an injury. Thus, despite the good will of the provincial agencies, they were unable to help much and the identification of excellent companies had to be largely based on word of mouth. Thus some excellent performers, particularly small or medium-sized companies, were undoubtedly missed.

Insurance companies should be obvious candidates for excellence in safety in a "white collar" environment. When approached, the chief medical officer of a large Canadian insurance company admitted that they had many injuries (e.g., repetitive strain injuries, vehicle accidents, slips and falls) but that they had neither comprehensive injury records nor a particularly strong safety program. The president of a large insurance company admitted to having no real quantitative knowledge of the injury situation in the company. One of the major banks was approached and found not to have good documentation of injuries. The author's knowledge of government office environments was that their safety records, if known, were mediocre. Thus no office example was included in the research.

DuPont Canada had long been known as one of the safest companies in Canada, holding the plant and industrial company records for time without a lost work injury. It ranked high in worldwide DuPont, for decades one of the world's safest companies. The author was familiar with this company, having been personally employed by it for many years.

The Shell group of companies is also well known for excellence in safety, and Shell Canada was found to be a worthy representative of the company and of the best in safety in the oil and gas industry.

It was desirable to include a company from forest products, Canada's largest industry. A search of the pulp and paper sector turned up no *companies* with world class safety. However, Abitibi-Consolidated's Fort Francis *mill*, for many years considered the safest large mill in Canada, was chosen as a research subject.

The Ontario Industrial Accident Prevention Association (IAPA) suggested the fourth Canadian company, S&C Electric, a medium-sized electrical equipment company that had effected an outstanding turn-around in its safety performance. It turned out to be an excellent choice.

These four Canadian companies are to be thanked and commended for their excellent cooperation in this study.

4.4 US COMPANIES

The original intent was to include one or two US companies in the research. The National Safety Council and the VPPPA[ii] were consulted and made several recommendations. Milliken and Company, one of the world's largest textile firms, is widely recognized for its excellence in quality and in safety. It appeared to be among the safest companies in North America, close to DuPont and catching up. Milliken is a private company that values its privacy highly, but it is also proud of its safety record and agreed to participate. Several other US companies were identified as potential research subjects but either failed to qualify when examined more closely or chose not to participate. Considerable time was spent with a very safe US utility, but in the end it decided not to participate.

4.5 OFFSHORE COMPANIES

The original intent of including at least one European and one Asian company proved very difficult to carry out.

In Japan, meticulous records are kept by JISHA.[iii] Hitachi was

---

[ii] The Voluntary Protection Programs Participants' Association (VPPPA) is an association made up of workplaces that have been inspected and certified as having a high standard in occupational health and safety.

[iii] The Japan Industrial Safety and Health Association.

already known to the author as a very safe company, and the JISHA records confirmed this. Hitachi sites have the best all-industry records for the length of time without a lost work injury. The Kanagawa plant chosen for study holds the top record—116 million exposure hours without a lost work injury! Some work was done with Hitachi, including a visit to Kanagawa while the author was in Japan on other business. However, the cost and language constraints made it impossible to thoroughly assess the management practices behind Hitachi's outstanding record without more extensive research than was possible in this project.

In a visit to Australia on other business, an unsuccessful attempt was made to identify a company with world class safety. Some that were suggested were subsidiaries of international companies that had already been considered elsewhere. Attempts by correspondence to identify a British company also failed.

It became clear that fieldwork in the specific country would be needed and that was well beyond the scope and resources of this research. As a result, this was a North American study, covering research on the management of occupational health and safety at five very safe companies—one US and four Canadian.

4.6   COMPANIES WITH VERY POOR SAFETY

The original research plan mentioned the possibility of investigating some companies with poor safety. With the decision to use the safety questionnaire as the main research tool, it became apparent that research with some companies with poor safety would provide a useful "calibration." It is a difficult task to persuade companies with very poor safety to allow a researcher to find out why they are so bad. In the end this was done at arm's length through provincial agencies. The research included only completion of the questionnaire by a cross section of the workforce, not a full investigation with site visits, focus groups, interviews, etc, as was done in the very safe companies.

The criterion for qualification as a poorly performing company was set at a LWIF greater than 7 for an average of at least 3 years, two to three times worse than the Ontario average and several orders of magnitude worse than that of the very safe companies. This research was only possible through the cooperation of the provincial WCBs, safety associations, and agencies. They readily identified companies with poor safety performance. The problem was one of confidentiality and willingness of the companies to participate. Many that were approached refused. In the end, five sets of data were obtained. These companies obviously have little time for safety, and thus the data cannot be considered as accurate as that from the very safe companies. The response rate was lower than with the safe companies. Nevertheless, unique and very useful data were obtained.

The poorly performing companies cannot be identified. They were in the vehicle manufacturing, packaging, steel, furniture, and wholesale grocery businesses. They ranged in size from about 250 people to several thousand. The sites surveyed in these companies employed from 80 to 500 people.

4.7   COMPANY ENVIRONMENT—CULTURE

The companies chosen for research represented a mixture of unionized and nonunion operations. Of the 16 plants surveyed in the 10 companies, 7 had unions and 9 did not, distributed almost equally between the very safe companies and those with very poor safety. (Unionization was not considered as a factor in the research.)

In selecting the companies for research, no attempt was made to choose companies with either good or bad organizational environments. In the author's opinion, the work environment and worker-management relations in the five very safe companies was generally fairly good to excellent. There was a protracted strike at the Abitibi-Consolidated mill in 1998, but it was part of a wider company-industry strike action. The environment and the worker-management relations at the companies with poor safety were not observed directly. In some of them, the written comments on the questionnaires indicated the environment to be poor.

5

RESEARCH METHODOLOGY

5.1 OUTLINE OF RESEARCH

The objectives and products described in the Introduction are repeated below.

1. To further develop the model of managing for outstanding safety that generalizes the concepts, beliefs, and practices used by very safe companies, in such a way that the knowledge could be transferred to other organizations. *(Specific Product: A Model of Managing for Outstanding Safety)*

2. To further develop the safety questionnaire as a quantitative method for assessing the state of safety management in a company—a means of *measuring* the extent that the concepts, beliefs, and practices of the model are present in an organization. *(Specific Product: A Safety Questionnaire)*

3. To conduct detailed surveys using the questionnaire at five companies with outstanding safety records and five with very poor safety. This would validate and help improve the model and the questionnaire and provide benchmark data. *(Specific Product: quantitative questionnaire data from five very safe companies and from five with very poor safety)*

4. To conduct on-site investigations, interviews, and focus groups at each of the five very safe companies (not for the companies with

poor safety). This would augment and support the quantitative questionnaire data and help identify the ways that the five companies achieve excellence in safety. *(Specific Products: supporting information to augment the questionnaire data; a case write-up for each of the five very safe companies)*

5. To develop a comprehensive theory of the management of safety and to test its validity through the research findings. *(Specific Products: a report, papers, and presentations that integrate the research findings into a comprehensive theory of safety management)*

The model and the questionnaire (products 1 and 2 above) are described in Chapters 2 and 3. Although they were available in a more-or-less complete form before the investigation of the 10 companies began,[i] the model and the questionnaire were improved through the research and through further consulting projects that were conducted at the time of the research.

This chapter describes the methodology for *using* the questionnaire at the 10 companies, the compilation of the questionnaire results, and the methods used in interviews and focus groups to develop supporting information.

5.1.1   Research at the Five Very Safe Companies

For the very safe companies, the goal was to undertake research with the corporate management and at one or preferably two units representative of the company.[ii] The research thus consisted of the following:

For the Corporate Management Group

1. Collect data on the company safety performance and general data on its business performance.
2. Have the safety questionnaire competed by the CEO and all of the top management team (6–15 people).

---

[i] Because there were some changes in the questionnaire during the research, there are fewer than 10 sets of data for some questions. A second, supplementary survey for questions not in their first survey was completed at two companies (3 questions for 1 company, 1 for the second).

[ii] The units researched at the five safe companies were plants, mills, or refineries, called "plants."

3. Interview the CEO, the vice president or senior manager of human resources, and the corporate safety manager, using a standard set of questions (about 1 to 1-1/2 hours for each interview).

For Each Unit or Plant Location

1. Collect data on the plant's safety performance and general data on its business performance.
2. Have the safety questionnaire completed by about 50 to 80 people including:
   - All of the plant's top management team (the plant manager and senior managers, usually 6–12 people)
   - A cross section of supervisory people (usually 7–10 managers, superintendents, supervisors, team leaders, foremen, etc.)
   - A cross section of working level people (35–50 people)
   - A cross section of professional people (usually 3 or 4 technical, accounting, etc.)
3. Visit the plant, tour the facilities, talk to a cross section of people.
4. Interview the plant manager, the human resources manager, and the safety manager, using a standard set of questions (about 1–2 hours each).
5. Conduct focus groups with 8 to 12 supervisors and with 8 to 12 working-level people.

To avoid an influence on the answers, the questionnaires were completed before the interviews or focus groups. With this sequence, a preliminary analysis of the questionnaire results was available to guide the interviews and focus group sessions and any anomalies or unusual results could be discussed. For example, if the questionnaire results indicated that recognition for safety achievement was at a high level, the reasons for this could be probed in focus groups and interviews with managers.

The results of the questionnaire survey and the case write-ups were reviewed with the very safe companies before preparation of the research report on which this book is based. This provided another opportunity to discuss their safety practices and to explore unusual findings.

5.1.2 Research at the Five Companies with Very Poor Safety

The research at the companies with poor safety was arranged for and managed by provincial government agencies, based on written instructions. It was restricted to completion of the questionnaire by plant personnel only. There were no site visits, focus groups, or interviews of leaders. Some information on each company and limited data on its safety performance record were obtained.

For Each Unit or Plant Location

1. Collect data on the plant's safety performance record.
2. Have the safety questionnaire completed by about 50 to 80 people including:
   - All of the unit's top management team (the plant manager and senior managers, usually 6–12 people)
   - A cross section of supervisory people (usually 7–10 managers, superintendents, supervisors, team leaders, foremen, etc.)
   - A cross section of working-level people (35–50)
   - A cross section of professional people (usually 3 or 4 technical, accounting, etc.)

5.2   COLLECTION OF COMPANY DATA—VERY SAFE COMPANIES

As part of the research, information on the five very safe companies was collected. These investigations started during the selection of the company for research and continued during the planning of the questionnaire survey and through the following discussions with company personnel. The intention was to develop an overall picture of the company, and particularly of the place of safety in it. Whenever possible, written reports, policies, and procedures were obtained.

The head offices and selected plants of each of the very safe companies were visited so that their facilities could be seen and informal discussions could be held with workers and others.

At the company level, the following information was sought:

General Information

1. General information on the company—its history, products and services, markets, geographical scope, market position, competitors, main locations
2. Organizational structure—its ownership, number of employees, structure

3. Vision and values—long-term goals, values, where safety fits in the company
4. Company performance—its long-term financial performance, growth, market position, quality achievements, and other such indicators

Safety Information

1. Organizational Structure for Safety
   - Organizational structure for safety management
   - How safety is related to human resources, environmental affairs, etc.
2. Values, Goals, and Policies
   - The safety vision, values, policies, and long-term goals
   - Annual safety theme and objectives
   - Annual and interim reports on safety performance
   - Systems for communicating safety information and performance data
   - Safety considerations in hiring, safety of contractors, and subsidiaries
   - Off-the-job safety policies
3. Statistical Data (corporate and unit statistics, preferably for 5–10 years)
   - Lost work injury frequency (LWIF)
   - Total recordable injury frequency (TRIF) or equivalent
   - First aid case frequency or equivalent
   - Off-the-job injury records (OTJIF)
   - Incident frequency
   - Other safety measurements particular to the company or the industry
   - Data on contractor safety, safety performance of subsidiaries
   - Benchmark data—within the corporation, compared with industry sector, compared with all industry (nationally and internationally)
   - Safety awards or citations
   - Nonquantitative safety achievements (e.g., previous year against objectives)
   - Employee attitude or perception surveys on safety

4. The Line Safety Organization
   - Management systems (e.g., senior management committees, departmental and business unit committees)
   - Line responsibility systems—safety objectives of line management, on-the-spot injury management systems (and relationship to doctors)
   - Workforce committees (e.g., employee, company-union)
   - Policies, procedures, and examples regarding disciplinary action
   - Contractor safety management systems
   - Systems for involvement of line managers
5. The Safety Resource System
   - Where it reports, how organized (central, plant, business unit, functional unit, networks)
   - The number of safety resource people, their background and training
   - Written mandate, annual objectives
   - Duties of resource people and involvement in management safety systems
6. The Main Safety Systems (who owns, maintains, and operates them, procedures, typical written reports of performance)
   - Statistics, responsibility for classifying injuries
   - Safety manual
   - Safety rules and practices
   - Injury investigation procedures
   - Modified duty and return-to-work systems
   - Safety meeting systems
   - Workplace safety audits
   - Preparation of interim reports on performance—annual audit to board, report to internal organization (management and employees)
   - Safety in hiring procedures
   - Off-the-job safety systems
   - Safety training—new employees, supervision, management, workforce
   - Equipment and process hazards review procedures
   - Preventative maintenance systems
   - Recognition and reward systems

Similar data were collected at the unit level.

5.3 QUESTIONNAIRE SURVEY PROCEDURES

The central element in the safety survey was the questionnaire, a technique that differentiates this method of assessment from others. Because of its importance in the research, the questionnaire procedures will be explained in some detail.

In administering the questionnaire, the following elements had to be determined:

- How many people, in what groups, and at what levels would be polled?
- How would they be contacted and persuaded to participate?
- What instructions were needed?
- How would individual confidentiality be maintained?

These considerations are illustrated with reference to the typical manufacturing organization shown in Fig. 5-1.

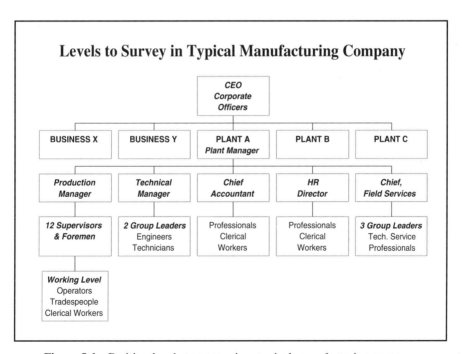

Figure 5-1 Position levels to survey in a typical manufacturing company.

5.3.1 Selecting the Levels, Grouping and Number of Respondents

The questionnaire was designed to poll four levels—management, supervision, working level, and professional. At the plant or unit level this included:

- **Management.** This is the management team—the plant or unit manager and those reporting directly to him or her—usually 6 to 12 people, most of whom would be line managers. In Fig. 5-1, the plant manager and the five managers that report to him/her would all fill out questionnaires.
- **Working Level.** This includes production and maintenance workers, clerical workers, and others in nonsupervisory, nonprofessional jobs. This is the largest group, and 35 to 50 people would be polled. The emphasis would be on those exposed to hazards—production, maintenance, warehouse, and laboratory workers.
- **Supervision.** "Supervision" is defined as those with responsibility for people and whose positions lie between top management and the working level—e.g., managers, superintendents, supervisors, foremen, and group leaders. In Fig. 5-1, this includes supervisors, foremen, and group leaders. Of the 17 shown, 7 to 10 would complete questionnaires. Most would be in line jobs, in this case in production and maintenance areas.[i]
- **Professional.** A sample of technical, financial, and human resource professionals who are not supervisory would be polled. However, the main emphasis would be on line positions—managers, supervisors and workers—those who have direct responsibility for the safety of others and those who face the most direct hazards. Thus in many cases professionals need not be included. In a plant of 500 people with 30 specialists, perhaps 3 or 4 might be polled.

The survey thus usually comprised 60 to 80 people from a 500-person workforce at the unit level.

Someone in the company, usually from the human resources staff, selected the individuals to be polled. They were asked to ensure that the selection was random. In particular, they were asked not to specifically select safety officials. Otherwise, the poll might be overweighted

---

[i] The questionnaire was set up to refer to traditional line organizations. In self-managing organizations, terms such as supervisor would likely not exist. Nevertheless, there would be people who provide guidance to the workers and handle many of the functions of supervisors.

with those most involved in safety matters—members of a Joint Health and Safety Committee, for example. However, they were asked to weight the selection toward operating groups, with fewer people from office positions.

In assessing the state of safety management of a *company*, enough units should be polled to give a good picture of the company as a whole. In the company in Fig. 5-1, a survey of two units might give a reasonable picture. All of the top management group would also complete the questionnaire—the president-CEO, the vice presidents, and other senior officers, usually 6 to 15 in number. In this research, the answers from executives were similar to those of the plant management group. In the assembly of the data, they were included with the plant management team as "management."

5.3.2 Definition of Specific Terms in the Questionnaire

Most of the terms in the questionnaire are self-explanatory or are defined in the questionnaire itself. A few needed to be defined for the specific situation.

The *organization* that the respondents were asked to think about when answering the questions was defined. Usually it was the plant or location where they worked, but in some cases respondents were asked to consider the state of safety in the company as a whole. People usually find it more difficult to answer questions about the company than about their specific unit. This is particularly true if the company has several locations or has its head office at a location other than the unit being assessed. The unit selected was the largest one to which the respondents could relate directly. Thus respondents in Plant A in Fig. 5-1 (a 500-person plant that is part of a 5-unit company) would think of that specific plant, not the company or their department within the plant.

The terms *management, supervision, working level,* and *professional* were also defined in the preface of the questionnaire, but more specific instructions for the particular organization were required in some cases. A key consideration was the internal integrity of the information. Everyone must be thinking about the same organization and the same practices. For example, the instructions for Plant A would be: "Think about the state of safety in *our plant*, Plant A, when you answer the questions. *Management* means the plant management team—the plant manager and the department managers; *supervision* means the supervisors, foremen, and group leaders; *working level* means the production, maintenance, clerical, and laboratory workers."

The CEO and corporate officers completed the same questionnaire and in most cases the terms meant the same thing. However, for them:

- The *organization* means the *whole company*, not a unit.
- *Working level* and *professionals* refer to positions across the company *at the plant or unit level*—the same groups referred to by the plant respondents.
- *Supervision* refers to the first level of supervision in the company—the supervisors, foremen, and group leaders. These are the same groups referred to by the plant respondents. Top management thus did not assess the top management teams of plants or other company units. This was so that all the results in the company referred to the same groups of people.
- For most questions, *management* has a generic meaning or it is defined in the question (e.g., line management). For question 2, *management* has a specific meaning for the corporate managers: it refers to the CEO and the top management team itself.

The definition of terms is shown in Fig. 5-2.

Definition of Terms in Questionnaire			
Term	*Corporate Management Respondents*	*Corporate Management & Plant Respondents*	*Plant Respondents*
Management	Themselves, The Corporate Management Team		
Management			The Plant Management Team
Supervision		The Supervisors, Managers, Foremen, Etc. Between The Plant Management Group & The Workers	
Professional		The Technical, Accounting & Other Professionals	
Working Level		The Working Level (Sometimes Also Clerical People, Etc.)	

Figure 5-2 Arrangement so that the plant & corporate people referred to the same groupings when completing the questionnaire.

5.3.3   Communication and Instructions for Completing the Questionnaire

The main company contacts (usually the president, vice president-human resources, plant manager, and safety manager) and the plants or departments involved were briefed on the purpose of the questionnaire. The company leaders were asked to communicate the background of the research to the workforce, telling them that the company supported the activity and urging them to complete the questionnaire. Beyond that, explanations to the respondents were kept to a minimum to avoid influencing the answers.

The main information communicated was:

· The questionnaire is part of a research project on safety management being undertaken by a professor-researcher from the University of Toronto. The intent is to improve the understanding of safety management by assessing the beliefs and practices in place in a cross section of companies.
· A cross section of your company—managers, supervisors, workers, and professionals—will be completing the same questionnaire.
· Please answer the questions as objectively as you can so that the answers will help reveal the true state of safety.
· Your individual answers will be kept completely confidential. Please do not sign your name. The individual data will not be seen by the company but will be turned over to the researcher for analysis. The answers will not be reported individually but only combined with the answers of others.
· The results for a company will not be reported separately in a public document without the permission of that company. They will be combined with the data from other companies and reported anonymously.
· Please do not show the questionnaire to or discuss the results with others unless they have also completed it. Others may complete the same questionnaire, and it is important that their views be as spontaneous as possible.

A written explanation tailored to the specific circumstances was included with the blank questionnaire distributed to the respondents.

5.3.4   Identification of Respondents

The company official who distributed the questionnaires was asked to complete the general identification section on the cover of the questionnaire. It listed the company location and the job category of the respondent. On the second page of the questionnaire, individuals were asked to identify their job category. The person distributing the questionnaires was asked to remind respondents to fill in this section. Otherwise, the answers might not be useable. This provided a second check on the job category of the respondents.

On the second page of the questionnaire, the respondents were asked to indicate whether they had an "official" position in safety and whether they were members of a union. An official capacity was defined as including such jobs as safety supervisor, safety steward in a union or member of a joint health and safety committee. Normal managerial or supervisory responsibilities for safety were not to be considered as official. This question was meant as a second check to ensure that the survey was not overweighted with safety officials but represented the workforce broadly. The person distributing the questionnaires in the company was also asked to avoid such a distortion, and the author was able to check this personally at most of the very safe companies.

The information on union membership provided another check on the person's job category but otherwise provided no useful insights. Of the 16 plants surveyed in the 10 companies, 7 had unions and 9 did not, and this was distributed almost equally between the very safe companies and those with very poor safety.

5.3.5   Ensuring Confidentiality

The people completing the questionnaire must feel confident that their answers are kept *completely confidential, particularly from their own management*. They were asked not to sign the questionnaire. As a further safeguard, an envelope was provided to forward the sealed questionnaire to a central location in the company.

5.3.6   Ensuring Good Coverage of the Organization

Ensuring sufficient coverage of the organization was sometimes a problem, with some people away and others not interested or concerned about the security of their answers. However, for most of the surveys of the very safe companies, a high proportion of the questionnaires were completed and returned (more than 85%). Returns were considerably lower among the companies with poor safety (50–90%).

5.3.7 Analysis of Results

The completed questionnaires for each company or company unit were sorted into the four job categories—management, supervision, professional, and working level. In most cases, the category of the individual could be identified from the information on the questionnaire. In the few cases in which the category could not be identified, the results were included in the overall database as "undesignated." The questionnaires were then examined for accuracy and consistency. Questionnaires with five or more incomplete or unusable questions were rejected. Thus the number of responses may differ from question to question for the same company. Then the data was entered into an Excel database and analyzed by category and response. For each company or company unit, there were five entries—four for the categories of management, supervision, professional, and workers and a final entry labeled *all* for the total of the answers for the four categories (plus any undesignated replies) for that item. In most cases, *all* is weighted toward the responses of workers, the largest group in the sample.

When the research included more than one company unit, all the individual results were added to get the *company* result (e.g., the result for workers was calculated as though there was one pool). The responses from executives were included with those from the senior managers in the plants because there were not enough differences to warrant separate analysis. The overall company result for management was thus calculated as a simple average of all the responses from corporate managers and managers in the unit or units.

The responses from professional people are listed separately in the Excel data tables in Appendix F. Their responses were not analyzed separately, however, but the data were included in the results for the total population ("all"). The focus of the research was on the positions that have the greatest potential for injury—the workers—and on those who are responsible for the safety of the operation—the managers and supervisors.

The combined responses for the safe companies (called "Safe Co. Avg.") and for the five companies with very poor safety (called "Unsafe Co. Avg.") are the simple average of the results for the five companies, not weighted by the number of respondents in each company. The "Best Result" and the "Worst Result" for each question were also selected. In multipart questions, the worst or best result for all parts of the question were usually from the same company but this was not always true. For example, company X may have had the worst result for workers but if all respondents were considered, company Y may have had the worst result.

At the end of the questionnaire, many of the respondents wrote in comments in the section left for that purpose. Some of the comments have been used as examples in the discussion of results.

The data for individual companies are not presented separately in the data summary in Appendix F because several companies requested that this not be done. For some questions, however, individual company answers are given, either in the discussion of that question in Chapter 6 or in the company write-ups in Chapter 7.

The data for all of the questions are tabulated in summary form in Appendix F. It should be noted that for confidentiality reasons, some potentially useful information was not included in the research report or this book (such as individual answers, unit data within companies, and separate data for corporate management).

5.4   INTERVIEWS AND FOCUS GROUPS

Interviews were held with corporate executives at four of the very safe companies (research at Abitibi-Consolidated was confined to the plant level) and with management at the plant level at all five companies.

The leaders interviewed were generally the following:

- At the corporate level, the president, the vice president or director of human resources, and the company safety officer. Nine corporate executives were interviewed, including the presidents of three of the companies
- At the plant level, the manager, the manager of human resources, and the safety officer. Fifteen managers were interviewed, including six plant managers

The interviews were based on a standard set of questions relating to the model of safety management. The same questionnaire and the same lists of questions were used for both corporate and plant people. (A copy of the questions is given in Appendix C.) Issues arising from the questionnaire results were also raised. Usually the interview took 1-1/2 to 2 hours and went well beyond the prepared questions. The interviews provided valuable insights on how the company managed safety. The interviews were recorded and used to help prepare the company write-ups.

The author had found in consulting that a good picture of the safety climate of the organization could be obtained by conducting focus groups with employees. This procedure was adopted partway through the research project. Formal focus groups were conducted with workers at three companies and with supervisory people at two. A variety of less formal meetings had been held earlier with employees of the other two companies.

A standard procedure was followed in the focus groups. To encourage openness, the groups met with only the researcher present. The participants were told that the purpose was to complement the data from the questionnaires and the interviews with leaders, that the results were for the researcher's use only, and that no names would be recorded. The researcher and sometimes one of the group recorded the

Company	Data Period	LWIF 5 Yr. Avg.	LWIF Range	TRIF 5 Yr. Avg.	TRIF Range
Safety Records of the Companies Studied					
Abitibi-Cons.	1993-97	0.11	0 to 0.4	2	0.90 to 4.9
DuPont Canada	1993-97	0.03	0 to 0.06	0.38	0.26 to 0.56
Milliken	1993-97	0.09	0.05 to 0.13	0.88	0.84 to 1.03
S&C Electric	1993-97	0.06	0 to 0.31	2.2	1.3 to 3.1
Shell	1993-97	0.12	0.08 to 0.17	0.65	0.50 to 0.83
Avg., Very Safe Companies	~	0.082	~	1.2	~
Company A	1993-95	7	na	na	na
Company B	1993-95	12.4	na	na	na
Company C	1993-95	14.3	na	na	na
Company D	1993-95	14.4	na	na	na
Company F	1997	53	na	na	na
Avg., Poorly Performning Co.	~	20	~	na	~
Five year averages for the safe companies, 1 to 3 year average for the unsafe companies					

Figure 5-3 The safety records for the five safe companies was 250 times better that that of the poorly performing (unsafe) companies.

points. The questions were intended to be neutral to encourage the participants to speak frankly. Below are samples of the questions. There were also specific questions that arose from the results of the questionnaire.

General Questions

· What do you think of the state of safety in this organization?
· Is it improving?

<div align="center">**Summary of Research at the Ten Companies**</div>								
Company	When Done	Corp. Manag. Quest.	CEO, Officer Interv.	Plants, Units Survey	Plant Manag. Interv.	Super. Focus Groups	Worker Focus Groups	Total Quest.
Abitibi-Cons.	1996	--	--	1	3	1	1	40
DuPont	1996-97	13	3	2	6	2	--	110
Milliken	1997-98	1*	1	2	2	--	1	129
S&C Electric	1997	5	3**	1	**	1	1	65
Shell	1997	6	3	1	3	1	1	55
Total		25	10	7	14	5	4	399
Company A	1995	--	--	2	--	--	--	45
Company B	1995-96	--	--	1	--	--	--	55
Company C	1995-96	--	--	2	--	--	--	40
Company D	1995-96	--	--	3	--	--	--	83
Company F	1998	--	--	1	--	--	--	30
Total		0	0	9	0	0	0	253
*GRAND TOTAL****		25	10	16	14	5	4	652
* Questionnaires were completed by 30 managers, professionals and workers who made up the Corporate Safety Committee. Only one was part of the corporate management group ** Corporate management & plant management located in one integrated unit *** The totals do not include 3 questionnaires from the safe companies and 15 from the companies with poor safety that were rejected								

Figure 5-4   Summary of the research done at the ten companies.

· What were the key reasons for achieving the present level of safety?
· Can you suggest an important way to continue to improve?

Specific Questions

· What ultimate level of safety can the organization get to?
· What is your view of management's commitment to safety?
· To what extent do you feel that all injuries can be prevented?
· How does involvement in safety activities work here?
· How are safety rules enforced?
· What is the role of the safety advisors?

5.5   SUMMARY OF COMPANIES AND RESEARCH UNDERTAKEN

Figures 5-3, 5-4, and 5-5 summarize the safety records of the 10 companies, the research undertaken at each and the number of questionnaires obtained for analysis.

Number of Questionnaires			
Job Category	Five Very Safe Companies	Five Companies With Very Poor Safety	Total
Corporate Officers	25	0	25
Other Managers	73	29	102
Total Managers	98	29	127
Supervisors	65	73	138
Professionals	41	14	55
Workers	191	116	307
Undesignated	4	21	25
Total Valid Data	399	253	652
Rejected	3	15	18
Total	402	268	670

Figure 5-5   The number and distribution of the questionnaires.

A summary of the safety records for the 10 companies is given in Fig. 5-3. More details are given in Chapter 7 in write-ups on the very safe companies.

The research is summarized in Fig. 5-4.

The distribution of the questionnaires is summarized in Fig. 5-5.

6

ANALYSIS OF THE QUESTIONNAIRE RESULTS

The model of safety management and the beliefs and practices that underpin it are discussed in Chapter 2. In Chapter 3, the questionnaire based on the model is described. The selection of the companies for research and the research methodology are reviewed in Chapters 4 and 5. In this chapter, the results from the questionnaire surveys completed at the five very safe and the five poorly performing companies are presented. Supporting information from the on-site investigations, interviews, and focus groups at the five very safe companies is woven into the discussion of the results for specific questions. (An integrated view of the safety management of each of the five very safe companies is given in Chapter 7.)

The validity of the model on which the questionnaire was based rests to a large extent on its derivation from the well-tested safety management methods of very safe companies. However, the questionnaire results also help verify the validity of the model. Because of this interrelationship, the concepts behind each question and the linkage to the model are reviewed before the results for that question are analyzed. All the questions relate closely to the model, and almost every question relates directly to a specific belief or practice.

The questionnaire was largely completed, tested, and refined before the research began. However, a few questions were added, revised, or removed during the research. Thus there are full sets of answers for most but not all of the questions. Some of the companies with poor safety were surveyed early in the research, so there are some data missing for them. The questions about safety meetings and return-to-work practices were added partway through the research. To have a good assessment of these two practices, resurveys were done for them at DuPont and S&C Electric. The data for these two practices were thus not from exactly the same people as the earlier surveys of these companies. There are data for these two questions from only three companies; otherwise, there is a complete set of data for all five of the very safe companies.

In most cases, enough people were surveyed to provide answers with reasonable statistical validity. The reader is cautioned, however, not to look for fine differences. A result of 10% answering "yes" in one company is not significantly different from a result of 15% in another company. Examples of statistical assessment are included in Appendix D.

The responses from corporate executives were included in the "managers" category with those from the senior managers at the plants; there were not enough differences to warrant segregation and separate analysis. The responses from the professional people are given in the data tables in Appendix F but were not analyzed separately. The main focus of the research was on the positions that have the greatest potential for injury—workers—and on those who are responsible for the safety of the operation—managers and supervisors.

The data for individual companies are not given because several companies requested that this not be done. For some specific questions, however, individual company answers are given, either here or in the company write-up in Chapter 7.

The data for the individual companies are weighted averages. For example, if questionnaires were completed by corporate management and at two plants, the number of managers in the three groups who responded that they were "deeply" involved was divided by the total number of managers who answered the question.

Most of the results are presented in the form shown in the examples that follow. Figure 6-1 gives typical data on involvement in safety activities for one of the companies with poor safety. The five possible answers to this question have been given short titles for ease in preparing the tables.

In Chapter 3, it is explained that four levels of insight can be gained from questionnaires of this type:

1. The absolute level of the answers yields important information.
2. Comparison among the answers from management, supervision, and workers yields important insights.

Involvement in Safety Activities One Unsafe Company					
Job Category	% Who Said They Are Involved ...				
	Deeply	Quite	Moderately	Not Much	Not At All
Managers	14	43	29	0	14
Supervisors	6	29	47	18	0
Workers	8	8	28	16	40
All	9	20	33	16	22

Figure 6-1 Involvement in safety activities was low in this company, particularly among workers (Q9).

3. The results from the subject company can be compared with those of other companies, particularly those with excellent and those with poor safety.
4. The results obtained at one time can be compared with those obtained later, to help determine whether improvement has occurred.[i]

In Fig. 6-1, two of these factors can be observed. The absolute level of the answers shows that involvement is not high in this company in any job category. Managers are more involved than are supervisors, and workers have the least involvement. Yet it is at the working level where injuries occur and where there is the greatest benefit from involvement.

Figure 6-2 gives comparative results for the same question. These data provide a third level of insight, an important one that places the individual company results in the context of the practices of the very safe companies. Involvement is much lower in the companies with poor safety. The results for the company shown in the previous example (Fig. 6-1) are at about the average for the companies with poor safety.

The individual company percentage data, such as that given in Fig. 6-2 for the best or worst company results, are the combined (thus weighted) responses from *all* the respondents in all the job categories. Sometimes the data refer to a specific job category; where this is so, it is noted. However, in calculating the averages for the very safe companies and for the poorly performing (or for short, the "unsafe")

---

[i] There are no data in this report dealing with the fourth level of insight, which requires a repeat survey. However, such data for a specific case are given in Reference 25.

Involvement in Safety Activities					
Company or Company Average	% Who Said They Are Involved ...				
	Deeply	Quite	Moder-ately	Not Much	Not At All
Best Result	46	28	18	5	3
Safe Co. Avg.	28	28	20	13	11
Unsafe Co. Avg.	8	21	21	21	29
Worst Result	3	13	7	30	47

Figure 6-2   Involvement is much higher in safe companies (Q9).

companies, the results were not weighted. Thus the figure of 28% of all respondents in the very safe companies that said they were deeply involved is the simple arithmetic average of the five results from the five companies. (Note that when the results from one of the very safe companies or one of the poorly performing companies are compared to the average, the average includes the results from that individual company as well as the results from the four others.)

Most of the analysis that follows is based on data summaries similar to these two examples. A few of the questions, particularly questions 1 and 2, required different handling. This is explained in the discussion of those questions.

Most references in the text refer to "the companies with very poor safety," or "the poorly performing companies." In the tables, the short form "Unsafe Co. Avg." is used to denote the average results for the companies with very poor safety. Similarly, the average results for the companies with very good safety are referred to in the tables as "Safe Co. Avg."

QUESTIONS 1 & 2: THE PRIORITY GIVEN TO SAFETY

The health and safety of people has first priority and must take precedence over the attainment of business objectives. (Belief 1)

The priority that the organization—its leaders and its people—gives to safety, in words and in deeds, is perhaps the most important single determinant of safety performance. It flows directly from the most important factor in the model—management commitment to excellence in safety. Thus the best indicator of the priority given to safety is the overall strength of the safety culture in the company and the thoroughness with which the important practices are implemented. The belief that safety must have overriding priority influences all aspects of safety. Although it appears simple, it has profound implications. The priority given to safety must be visible in all the actions of the company, particularly in the behaviour of management.

Giving overriding priority to safety does not imply that costs, quality, customer service, production volume, and other business parameters are not critically important. They are the lifeblood of successful companies. Rather, this belief means that in *any* case of conflict between safety and other objectives, safety is given overriding priority. Then the organization must "go back to the drawing board" to resolve the conflict and find a way to deliver excellence in all parameters. "All the balls must be juggled."

There is long experience behind the phrasing of this value. One organization with good but not excellent safety worked to the credo, "We have no higher value than safety." Although this may seem the same, it is critically different from giving safety overriding priority. Costs, quality, and volume are at the top of everyone's mind every day. Unless *overriding* priority is given to safety, it will likely drift to a lower priority in the face of pressure for business performance, as it had in the organization mentioned above.

Safety, integrity, and quality can be envisaged as being in a different dimension from other business parameters. They are "the way we must do everything," without compromise. Just as we will not cheat to improve profitability or trade off product quality for short-term cost improvement, we will not compromise safety to increase production volume. Safety, dealing with human life, must have overriding priority.

Whatever words are used, it is critical that people truly *believe* that the organization puts the highest value on the health and safety of its people, higher than the value it puts on monetary factors. Perception is all-important; if the organization is perceived to compromise safety, cynicism will quickly develop. Management that *does not* give high priority to safety is giving an unmistakable signal to its employees. Leaders in the safest companies know that unswerving commitment to the priority of safety is critical. If this value is not explained clearly and meticulously observed in practice, employees will realize that their safety is really given lower value and will act accordingly.

Some leaders are reluctant to give unequivocal first priority to safety, thinking that it will take away from the priority employees give to business factors. The very safe companies know that there is no conflict between outstanding safety and excellence in business, that they are mutually supportive.

A corollary of this belief is the understanding that the safe way to do things is the best and most effective way in the long run. A good safety practice must meet the double hurdle of defining the safe way to do the job and the economical, effective way.

Two related questions were constructed to help judge the priority given to safety. One addresses the respondent's personal priorities, and one addresses the perception of the priorities of other groups in the organization.

The Results (Question 1): The Priority *Individuals* Give to Safety

In this question, the respondents were asked to rank their *personal* priority among quality, costs, production volume, and safety.

1. Indicate the **priority you personally give** to the following items. Rank in order from 1 to 4, with the item you think is the most important marked 1 and the least important marked 4.

Item	Your Priority
A. Quality, customer focus	
B. Costs, efficiency	
C. Production volume	
D. Safety	

The respondents were asked not to give any of the factors equal rating; if they did, their answers were discarded. This gave some people problems, particularly in companies with poor safety. Some respondents wrote in that the question was not valid because all of the factors are important and none should be given precedence. This indicated a superficial understanding of the belief. Figure 6-3 gives the results from question 1.

In the very safe companies almost all respondents said that they give first priority to safety, and the results were fairly uniform. In companies with poor safety the priority given to safety was much lower

The Priority Individuals Give to Safety				
	% Who Ranked Safety First			
	Manag.	Super.	Work.	All
Best Result	97	90	92	94
Safe Co. Avg.	84	79	91	83
Unsafe Co. Avg.	46	64	65	62
Worst Result	44	58	53	56

Figure 6-3   In the very safe companies, individuals give much higher priority to safety (Q1).

and less uniform. In the "All" (all respondents) category, none of the responses from the five companies with poor safety was as high as the lowest of the responses from the companies with very good safety. Generally, workers and supervisors in the companies with poor safety said that they give higher priority to safety than their leaders reported.

Comments written in on the questionnaire by people from the companies with poor safety supported the questionnaire data.

The Results (Question 2): The Priority People Think *Others* Give to Safety

In question 1, respondents were asked to report the priority they, as *individuals*, give to safety. Although this is an important question and the results correlate with safety performance, there is a strong element of subjectivity in the answers. Much more important is the view that individuals have of the priority *others* give to safety, and particularly the view they have of management's priority. Thus, in question 2, the respondents were asked to assess how they think others—managers, supervisors, and workers as groups—rank the priority of the factors.

2. Indicate where **you think others** in your organization rank the same items. For example, give your opinion of the priority that you think supervision as a group gives to the item. Rank in order of priority from 1 to 4 as in question 1.

Item	Priority of Management	Priority of Supervision	Priority of Workers
A. Quality, customer focus			
B. Costs, efficiency			
C. Production volume			
D. Safety			

The answers to this question were revealing. Where safety was very good most respondents said that they and other groups give safety high priority. Where safety was poor the answers indicated a much lower priority given to safety, and there was almost always a split opinion: managers thought that they give priority to safety but workers said that their managers do not and vice versa.

Figure 6-4 shows the results for S&C Electric, one of the very safe companies.

In Fig. 6-4, the column "Selves" repeats the answers from question 1, where 82% of the managers at S&C Electric said that they individually give safety first priority. Ninety-one percent of the managers thought managers *as a group* give safety first priority, eighty-two percent thought supervisors give safety first priority, and ninety-one percent thought workers rank safety first. Supervisors were more critical of themselves, both individually and when they considered super-

The Priority People Think Others Give to Safety (S&C Electric)					
	% Who Thought Others Rank Safety First				
	Selves	Man.	Sup.	Work.	Avg.
Managers	82	91	82	91	88
Supervisors	62	77	62	100	79
Workers	91	91	82	91	88
All	75	81	72	86	80

Figure 6-4   At S&C Electric, people trust that others give high priority to safety (Q2).

The Priority People Think Others Give Safety Company with Poor Safety					
	% Who Thought Others Rank Safety First				
	Selves	Man.	Sup.	Work.	Avg.
Managers	67	33	33	0	22
Supervisors	58	25	50	23	33
Workers	53	29	24	29	27
All	56	28	35	22	28

Figure 6-5   At this unsafe company, few thought that others give high priority to safety (Q2).

visors as a group. They perceived that managers and workers give higher priority to safety than do supervisors. Workers gave managers, supervisors, and their own worker group high marks. At the bottom right of the figure is the combined average view of all respondents: 80% thought that the people in the company as a whole rank safety first.[ii] These very good overall results would have been even better had the relatively large number of professionals not given middle-of-the-road answers. (The company recognized that professionals were not as much involved in safety as they should be.)

Everyone in a focus group of a dozen workers at S&C Electric agreed that safety is given first priority. They agreed unanimously that management is committed to excellence in safety and expects workers to stop immediately anything that poses an undue risk. The workers were unusually open and confident of their ability to manage their own safety and sure of the support of the company, refusing to indicate any cynicism. They pointed out that doing the job safely was part of "doing it right the first time" and would save effort in the long run.

Contrast the results from that very safe company with those from one with poor safety shown in Fig. 6-5.

Sixty-seven percent of the managers polled in this company claimed that they *individually* give first priority to safety, but only thirty-three percent believed that management as a group does. Workers individually said that they give quite high priority to safety (53%) but have a

---

[ii] The numbers in this column, for this question only, are *averages*. In the figures for all other questions, the numbers in this column, labeled "All" are the *combined* results of the answers of the respondents.

lower opinion (29%) of the priority of their peers. They thought that managers and supervisors give very low priority to safety. No one gave anyone else much credit for valuing safety. No wonder the safety performance was so poor—more than 100 times worse than in the very safe companies. The comments from people in the companies with poor safety were in the same vein. For example: "I think safety is the most important thing about work. The supervisors just care about production volume and profits. They would rather get you hurt as long as the job gets completed on time! They don't take the blame for injuries. They say we are careless." A typical comment from a worker in one of the worst performing companies was: "Safety in our organization takes a back seat to productivity. If they did not have to worry about safety issues, they wouldn't. Productivity is everything, safety is nothing."

The priority actually given to safety in the companies with very poor safety is probably lower than indicated by the distressing answers, given their very bad injury frequencies. People in these companies obviously do not understand the meaning of giving first priority to safety.

People in the safe companies, as illustrated by the data in Fig. 6-4 for S&C Electric, did not claim that they as individuals give higher priority to safety than their managers and co-workers. The average for "selves" was 75%, for others 80%. In contrast, in the companies with poor safety, people said that they give higher priority than do others. In the company illustrated in Fig. 6-5, 54% said that they give first priority to safety but believed that only 27% of others did. In the author's consulting experience, this kind of dichotomy is pervasive in the results for companies with mediocre or poor safety.

The data in Fig. 6-6 shows the spread between the best and the worst in this most important measure of commitment to safety.

The answers of the managers and supervisors were averaged. Although there was some variation in the data from the five very safe companies, their answers were much higher and more uniform than those of the companies with poor safety. In the best result, almost all the managers and supervisors (96%) said that workers give safety first priority and almost all the workers (87%) said managers and supervisors give safety first priority.

The average result for the safe companies would have been considerably higher (83–75% compared to 81–66%) had the result for Abitibi-Consolidated not been included (see write-up on this company in Chapter 7).

Among the companies with poor safety, the results were more erratic. On average, only 39% of managers and supervisors thought

The Priority People Think Others Give to Safety					
	% Who Thought Others Rank Safety First				
	Best Result	Safe Co. Avg.	Unsafe Co. Avg.	Worst Results	
Managers' & Supervisors' View of Workers' Priority	96	81	39	12	41
Workers' View of Managers' & Supervisors' Priority	87	66	19	27	8

Figure 6-6 People in safe companies have more confidence in their co-workers than in companies with poor safety (Q2).

that workers give safety first priority but the range of opinion—from 12% to 72%—was wide. Workers in all five companies had a poor opinion of the priority that management gives to safety. Only 19% thought that their management gives first priority to safety. The worst results were very striking. Two are cited. It would be difficult to imagine attaining good safety where only 8% of workers thought management gives safety first priority or in the other company, where only 12% of the managers thought workers give safety first priority.

Comments on Questions 1 and 2

In the very safe companies, the priority given to safety is viewed as a critical issue. They believe that one of the strongest factors that predicts safety performance is *the perception of workers about the commitment of their management to safety.* The results shown here bear out their view. The data in Fig. 6-6 indicate how well this question differentiates the safe from the unsafe companies. The answers to question 2 give one of the very best indicators of management commitment and are thus an excellent predictor of safety performance. This is one of the most important questions in the survey.

QUESTION 3: THE BELIEF THAT ALL INJURIES CAN BE PREVENTED

All injuries and occupational illnesses can be prevented. Safety can be managed and self-managed. (Belief 2)

The belief that all injuries can be prevented is one of the most fundamental values. It has implications for all aspects of safety. If all injuries *can* be prevented, it is incumbent on leaders to see that they are. This also means that injuries can never be blamed simply on worker negligence, thus supporting the belief about the responsibility of line management (Belief 7).

Few leaders believe that *all* injuries can be prevented, and the same is true of most workers. Yet, fully understood and used, this belief has a powerful influence on safety. Even in organizations with good safety, all the exceptions emerge when this belief is discussed. "How can you prevent injuries from an earthquake?" "What about the driver that runs the red light and crashes into you?" There are situations where avoidance of injury seems very difficult. But in a practical sense, the safest companies have gone a long way towards proving the adage. The best statistical performance among the five safe companies in this research project was that of DuPont Canada. The company recorded five lost work injuries with 3500 employees in the five years from 1993 through 1997. Had it experienced the average injury rate for manufacturing in Ontario, it would have had more than 500. Thus it could be said that more than 99% of the potential lost work injuries were prevented. The safe companies observe that the few injuries that are sustained could have been prevented. They do not use the term *accident*, with its connotation of an uncontrollable event. Instead, they refer to injuries and incidents. As the belief says, "*Safety can be managed . . .*".

The difference between *all* injuries being preventable and *almost all* may seem unimportant. Yet it is a critical difference, one that indicates a different view of safety management. Which are the injuries that *cannot* be prevented? An injury can be explained: "What can we do about a worker who didn't pay attention and was injured? That was one that couldn't be prevented, an unfortunate accident, for which I as the manager can't take responsibility." If the organization truly believes that all injuries can be prevented, an injury cannot be shrugged off solely as the result of a careless worker. The worker is responsible and accountable for his or her safety. But so is the supervisor, who must see that safety awareness is developed, that all steps are taken to see that injuries do not occur. So are the manager and the president, who are responsible for creating an environment in which people are not injured.

Who wants to work for a manager who thinks that only *most* injuries can be prevented? Would he or she see that everything was done to prevent an injury *to me*? Or shelter behind the excuse that my injury was one that could not be prevented?

The belief includes the phrase "safety can be self-managed." Companies that have been successful with self-directed teams have learned that, although somewhat different techniques are required, safety can still be managed to eliminate injuries. Through developing "ownership" in the minds of all employees, the safety culture in self-managed organizations can become even stronger than in more traditional structures.[iii]

The Concept of the Extent of the Hazard

Conventional thinking holds that it is more difficult, even impossible, to prevent injuries in hazardous workplaces. "It's easy to have a good record in the clean, automated chemical industry, but in the steel industry (or pulp and paper or whatever) there are far greater hazards, so we couldn't expect to achieve the same level of safety." This thinking is contrary to the belief that all injuries can be prevented. If the nature of the work is accepted as an excuse for injuries, we are on a truly slippery slope. Some companies in hazardous industries have proven that, although it may be more difficult to eliminate injuries in their operations, it can be done. The safest companies reject the hypothesis that safety performance depends on the specific hazard.

The other side of this coin is the observation that some companies with ostensibly minor hazards sustain a relatively high level of injuries. In this research, a number of companies whose operations mainly involve office work were found to have poor safety records. Typical was a Canadian insurance company with several thousand people. The expectation was that its involvement with risk and its largely office environment would lead to excellence in safety. Not so: The chief medical officer was unable to cite injury statistics: they didn't keep them in a consistent form. Do they have injuries? Indeed they do, replied the medical officer, lots of them—slips and falls, car accidents, repetitive strain injuries.

This contrasting experience of hazardous operations with excellent safety and companies with few visible hazards yet with high injury rates is not as surprising as it might seem. The central thesis of this book is that beliefs and attitudes, not physical things, are the fundamental determinants of the level of excellence in safety.

[iii] See Ref. 3 for a discussion of safety in a self-managed environment.

The Results (Question 3)

3. To what extent can **injuries be prevented**? Check the answer that represents your personal belief.
   - ☐₁ All can be prevented
   - ☐₂ Almost all can be prevented
   - ☐₃ Many can be prevented
   - ☐₄ Some can be prevented
   - ☐₅ Few can be prevented

The results are summarized in Fig. 6-7. There was considerable scatter in the results, even among the safe companies, reflecting the range of opinion on this belief.

In DuPont Canada, the company with the best result, three-quarters of all respondents and almost all managers and supervisors said they believed that all injuries can be prevented. However, even though this value has been an article of faith in the company for generations and is regularly reinforced, not everyone subscribes completely to it, although all of the 110 respondents checked off either "all" or "almost all." In the companies with very poor safety, few respondents said that they believed that all injuries can be prevented, and in the company that recorded the worst result, almost no one did. Not surprisingly, DuPont's safety record is more than 200 times better in LWIF than that of the company with the worst result.

Workers generally reported a lower level of belief that all injuries can be prevented, although in the worst result, a few workers but no managers or supervisors did check off "all." This is a sophisticated and

The Belief That Injuries Are Preventable			
	% Who Believed *All* Injuries Can Be Prevented		
	Man.	Super.	Work.	All
Best Result	86	88	64	75
Safe Co. Avg.	67	73	46	57
Unsafe Co. Avg.	25	23	15	20
Worst Result.	0	0	11	9

Figure 6-7 The belief that all injuries can be prevented is held much more strongly in the safe companies (Q3).

somewhat abstract belief, so if leaders have doubts, it is not surprising that the rest of their organizations does as well.

The manager of Milliken's Monarch, SC textile plant subscribed completely to this belief. He felt that getting people to believe that all injuries are preventable was the biggest barrier to their improvement in safety. He said, "It is also the biggest barrier to us improving further. Not just what people say but that they truly believe it and act on it."

Comments on Question 3

This question deals with a relatively abstract belief but one with important implications for safety. The results discriminate well between the very safe and the poorly performing companies. Like many of the questions, this one is very useful in a consulting situation: Discussion of the specific results of the survey with employees affords an opportunity for a full exploration of the importance of this value.

QUESTIONS 4 AND 5: THE INTERACTION BETWEEN BUSINESS AND SAFETY

Excellence in safety is compatible with excellence in other business parameters such as quality, productivity, and profitability; they are mutually supportive. Safe, healthy employees have a positive impact on all operations. They have a positive effect on customers and enhance credibility in the marketplace and in the community. (Belief 3)

Most leaders will espouse the belief that excellence in safety has a positive effect on business performance. But many do not really believe it; instead, cost-benefit trade-off thinking leads to compromises in safety.

The first part of this belief relates safety performance to success in measurable business parameters such as costs, productivity, quality, and profitability. The second part refers to the interaction between safety and factors such as employee morale that are less tangible but nonetheless have a great effect on business performance.

It is difficult to give safety first priority unless the leaders believe that, at least in the long run, excellence in safety will yield economic as well as human dividends. It is commonly believed that, beyond achieving fair safety, further improvement will cost more than it will yield in benefits—the so-called cost-benefit trade-off in safety discussed in Chapter 2. This disabling view is one of the reasons that leaders do not drive persistently for excellence in safety and thus why the frequency of injuries is high.

In contrast, the safest companies believe that outstanding safety has a substantial net benefit, not a cost.[iv] They understand that although there are direct savings from excellent safety, the greatest benefits are in the intangibles—in the effect on teaming and morale and in the spin-off for other things like quality and environmental protection. The quality of skill, knowledge, and team cooperation needed to deliver excellence in safety is reflected in all other activities. These companies have safety so deeply built into their cultures that deterioration in safety performance is seen as an early warning of a more general malaise in the organization. In the spirit of continuous improvement, they believe that as they get better in safety they will continue to find ways to improve even further. The analogy to quality is apt. A generation ago, quality was considered a cost: "Meet the customers' specifications but don't go too much further or it will cost you and will not benefit the customer." The Japanese quality revolution changed this thinking. We now say "Quality is Free," even quality 1000 times or more better than it was 20 years ago. Some are achieving 6-sigma levels in defects (defects per million in single figures). Today's corporate leaders realize the enormous benefits of excellence in quality. Surely, in the same context we can say "Safety is Free."

This is not to deny the need for an up-front investment to build excellence in safety. But in the long run it will pay off in many ways, perhaps the least of which is the measurable direct monetary benefit.

It is not a good idea to be simplistic about the economic benefits of excellent safety. If safety is poor, improvement can be shown to result in direct dollar savings in the costs of workers' compensation, in lost production, and so on. However, if safety is at a "fair" level, what is the economic justification for improving further? The benefits cannot be easily quantified, and astute CEOs will ask for proof of what is in it for them. It is not good enough to cite the slogan that safety pays. The best evidence is the excellent business record of the very safe companies and the deep belief of their management that outstanding safety pays off in direct and particularly in indirect ways.

The belief about the relationship between safety and business excellence is one of the most important. Unless management holds to and lives to this belief and convinces the workforce that it is a central

---

[iv] Even safe companies have difficulty with this concept and only come to believe it when they are successful in both safety and business results. See Ref. 3 for a description of managing this issue at DuPont Canada.

company value, it will be difficult to attain excellence in safety. Despite what they might say, if leaders really believe that attention to safety takes away from the ability to deliver business results, they will not strive for excellence. If they fail to understand the great intangible benefits of safety excellence in the parallel drive for outstanding quality, if they do not perceive the great benefits in employee motivation, in community and customer respect, they will not have the passion for safety needed to inspire their organization.

The Concept of Risk

Allied to the issue of the cost-benefit trade-off in safety is the concept of risk. The conventional belief is that as productivity is improved, the risk of injury increases (people will be busier). The extent of the increased risk must be assessed to see that it is acceptable. This is not the way to think about safety. It is not acceptable to increase the risk of injury. Instead, the risk should be *reduced* as we improve business performance. As we improve quality, we reduce the risk of defects and lower costs. The same applies to safety. This prevalent view of risk is a subset of the view that safety costs. It must be talked out—and eliminated—to achieve high levels of safety performance.

Doubts about the benefits of safety excellence can emerge in a number of places in the answers to the questionnaire. Unless leaders believe that excellence in safety brings business as well as human benefits, they will not give safety high priority. This will show up in the answers to questions 1 and 2 about the priority given to safety as well as in the direct questions 4 and 5.

The two questions that follow attempt to get a direct reading on the beliefs that people hold about the interaction between safety and business. They do not relate to *action* that is being taken on the beliefs.

The Results (Question 4): The Effect of a Drive for Safety on Business Performance

In question 4 the respondents were asked their opinion on the effect a drive to improve safety would have on the ability to achieve excellence in business results.

4. Indicate how you think a drive (a strong, long-term effort) for excellence in safety would affect **the ability to achieve excellence in other areas**, such as quality, productivity, costs, and profits. Check only one answer.

The safety effort will:

$\square_1$ Be very helpful in achieving other business objectives.

$\square_2$ Provide some positive assistance in reaching other business objectives.

$\square_3$ Have neither a positive nor a negative effect on attaining other business objectives.

$\square_4$ Tend to make it more difficult to achieve other business objectives.

$\square_5$ Substantially weaken the ability to achieve other business objectives.

There was a definite difference between the answers from the very safe companies and those from the poorly performing companies (see Fig. 6-8). Almost everyone in the safe companies said that they believe that safety has a positive influence. In the companies with poor safety, a considerable number of respondents doubted the positive influence of good safety.

Figure 6-9 gives another view of the answers to this question. Here the proportions of respondents from the four job categories who said that they felt the drive for safety would be "very helpful" are listed. (Note that the "worst result" cited in Fig. 6-9 is for the company whose respondents gave the lowest responses for the "very helpful" answer, a different company than for the worst *overall* result cited in Fig. 6-8.)

Here also a significant separation can be seen. In all cases, but particularly in the companies with poor safety, workers doubted the benefits of excellence in safety more than did managers. This has been

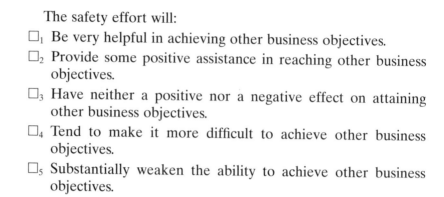

The Effect of a Drive for Safety on Business			
	% Who Said Safety Drive Affects Business Performance in Following Way ...		
	Very Helpful	Positive	Neutral	Makes Difficult, Weakens
Best Result	73	26	0	1
Safe Co. Avg.	67	29	1	3
Unsafe Co. Avg.	32	40	15	13
Worst Result	24	42	12	22

Figure 6-8   People in safe companies have a positive view of the effect of safety on business performance (Q4).

The Effect of a Drive for Safety on Business				
	% Who Said Safety Drive Is Very Helpful			
	Man	Sup	Work	All
Best Result	86	80	68	73
Safe Co. Avg.	76	71	62	67
Unsafe Co. Avg.	56	32	24	32
Worst Result	33	21	12	16

Figure 6-9 Managers are more convinced than supervisors or workers that safety helps business performance (Q4).

borne out in focus groups of workers in consulting projects. Workers in organizations with poor safety say, "It is just common sense that if we work harder on safety, something else will suffer." This is true in the short term, possibly, but the examples of the very safe companies (in this research, for example) have proven that excellence in safety can be combined with excellent business results.

The Results (Question 5): The Cost-Benefit Break Point in Safety

This question addresses the interaction between safety and business from a different angle. It tries to determine whether the respondents believe that there is a tangible limit in safety performance beyond which the effort to improve further will start to cost more than it benefits. If people believe that there is a limit, they will be less likely to strive for excellence. This is much like the concepts of product quality. The pioneers of the quality revolution did not accept the conventional thinking about the cost-benefit balance in quality and were thus motivated to drive quality to levels previously unheard of. Yet costs continued to come down.

5. At what level of safety performance do you think the effort to improve safety starts to **cost more than it yields in "economic benefits"?** Economic benefits means savings from the reduction in the costs of injuries, lost working time, loss of material, etc., and the indirect economic benefits that good safety brings in better morale, improved production, better product quality, etc. Check only one answer.

☐₁ Within reason there is no limit. *Exceptional* safety performance returns more in economic benefits than it costs.

☐₂ Once safety performance is at the *excellent* level, further improvement will cost more than the economic benefits it delivers.

☐₃ Once *good* (well above average) safety performance is reached, further improvement will cost more than the economic benefits it delivers.

☐₄ Once *average* safety performance is reached, further improvement will cost more than the economic benefits it delivers.

☐₅ Safety programs are a net cost. Improving safety always costs more than it benefits.

The results given in Fig. 6-10 show a difference between the answers from respondents in the very safe companies and those in companies with poor safety, but it is not as great a difference as one might expect. Also, there was some overlap,[v] with the result from one of the safe companies a little lower than that of two of the poorly performing companies. Apparently, some people in that safe company believe that to improve their safety beyond the excellent level might cost more than it would deliver in benefits. On the other hand, if people in the companies with poor safety truly believe that they would continue to gain economic benefits as their safety is improved (from truly bad levels in all cases), why would they not work harder at it?

The Cost-Benefit Break Point			
	% Who Said That at This Level Safety Starts to Cost More Than It Benefits ...		
	No Limit	**Excellent**	**Good or Average**	**Always Net Cost**
Best Result	92	7	1	0
Safe Co. Avg.	77	21	2	0
Unsafe Co. Avg.	67	12	12	9
Worst Result	47	23	20	10

Figure 6-10 People in the safest companies think that even at exceptional levels, further improvement will be beneficial (Q5).

[v] One of only three questions where there was any overlap between the (combined) answers from any of the safe companies and any of the companies with poor safety.

Comments on Questions 4 and 5

The belief about the interaction between safety and business success is one of the most important ones. Unless management holds this belief and convinces the organization of its power, it is difficult to attain excellence in safety. Even if safety is good, if the leaders feel that their organization is close to a cost-benefit balance, this will be one of the main barriers to achieving outstanding safety.

Many of the questions in the questionnaire reveal the views of the respondents about the place of safety in the organization—in the perception of the priority given to safety, for example (questions 1 and 2), and in the attention given to key practices. Questions 4 and 5 attempted to dig deeper to try to understand why priority was or was not given to safety. During the research a number of questions were tried in an attempt to find one that clearly measured the presence of the belief about the business-safety interaction in the culture of the organization. None was completely satisfactory.

The results from question 4 show a definite correlation: respondents from the companies with very good safety were more likely to view a drive for excellence in safety as positive for business results.

The results from question 5 do not provide as clear a picture. In the very safe companies, many but not all respondents said that they believe that even at excellent levels of safety, further improvement will provide a net benefit. However, even in the poorly performing companies, a surprisingly high proportion of people holds the same belief. Perhaps people answer this question without thoroughly considering the contradiction between beliefs and actions. They may say they do not believe that there *should be* a conflict between safety and business, but they may see such a conflict in their workplace, or even, as leaders, accept it. This lack of true dedication to striving for excellence in safety is evidenced in the answers to other questions, for example, in the questions about the priority given to safety. There the question is not about a belief but about an observation of the real situation.

QUESTION 6: THE EXTENT TO WHICH SAFETY IS BUILT IN

Like quality, safety must be made an integral part of every job. "Do it right the first time." (Belief 4)

The belief that safety must be made an integral part of every job is analogous to the belief that has helped drive quality to new levels. It proposes that to prevent injuries we must build safety into all that we do rather than try to add it on. This belief is based on the concept that safety, like quality and integrity, is in a different dimension than other business parameters, a modifier of everything we do. We do everything safely, with high quality and integrity. Safety becomes an integral part of management strategy, of annual corporate objectives, of dealing with suppliers and customers, of the development and implementation of new processes, of facility design, of hiring processes—of all aspects of the organization's operations. Safety training is built into job and task training.

The power of this concept has been proven in quality management. Major progress in quality was impossible until this concept was accepted. As in quality, doing it right the first time is more a mind-set issue than one of equipment and procedures. Everyone must "Take Two"[vi] to consider safety before any task is undertaken.

Phillip Crosby recognized this: "Everything safety is about relates to the absolutes of quality management. Safety and quality have the same basic structure of laying out a system and managing it such as to make things happen properly" (26).

The Results (Question 6)

6. To what extent is safety "designed in" to the facilities and **"built in"** to the operating practices in your organization? Built in means that safety is considered as an integral part of the design of equipment, of the development of operating practices, and of job training, not something that is added on later. Check only one answer.

   $\Box_1$ Thoroughly built in
   $\Box_2$ Substantially built in
   $\Box_3$ Some integration
   $\Box_4$ Little integration; mainly added later
   $\Box_5$ Not integrated at all; added later

The results in Fig. 6-11 show a clear difference between the opinions of the respondents from the very safe companies and the one result from a company with poor safety (only one was polled). In the best

---

[vi] "Take Two" is a slogan used by DuPont to remind everyone, particularly workers at the commencement of a task, to take 2 minutes (symbolically and actually) to think through the safety implications of what they are undertaking. The same slogan is useful in quality.

The Extent Safety Is Built In			
	% Who Said Safety Is Built In ...		
	Thoroughly	Substantially	Some Integration	Little or No Integration
Best Result	54	46	0	0
Safe Co. Avg.	42	47	8	3
Unsafe Co. Avg.	13	17	36	34
Worst Result	13	17	36	34

Figure 6-11   In safe companies, safety is more strongly built in than in companies with poor safety (Q6).

result, all the respondents said that safety was thoroughly or substantially built in.

Comments on Question 6

This question was added during the research and while results were obtained for all the very safe companies, only one company with poor safety was polled. Thus further work is needed to determine the validity of this question in predicting safety.

QUESTION 7: THE PRESENCE AND INFLUENCE OF SAFETY VALUES

Top management must be committed to excellence and drive the agenda by establishing a vision, values, and goals, by seeing that all line managers have safety improvement objectives; by auditing performance; and by visible personal involvement. (Belief 6)

This belief about safety management defines the main responsibilities of the CEO (and top leaders). The implications of this belief show up in the strength of the safety culture, in the priority given to safety, and in the practices of the organization. Thus it is tested by many questions in the questionnaire, not just this one about values. However, values provide the central focus and framework for safety in the very safe companies.

Well-intentioned leaders sometimes believe that they are committed to safety, yet their organizations do not deliver excellent safety results.

The message in this belief is that *words are not enough*. What CEO would delegate improvement in profitability to the chief financial officer and then go on with other things? The leaders must "bulldog" this issue as other important ones, leading the development of a vision, wrestling with the development and promulgation of a set of values, and establishing safety goals. The leaders must ensure that the goals are translated into specific, measurable objectives that represent both a step-change (if safety is mediocre) and continuous improvement. They must see that all managers have specific objectives that relate to the corporate objectives. They must audit the attainment of the objectives and, if improvement is not forthcoming, see that other approaches are developed that do work.

The leader must be visibly involved. In most of the very safe companies, the key leader (e.g., the president, the mill manager) personally chairs a management safety committee. This sets the tone for line management to take direct responsibility. True personal involvement is not just supportive speeches and attendance at safety affairs, although these actions are useful. Far more important is the hard planning required to put a change in beliefs, practices, and culture in place and the insistence on results.

In the safest companies, management by principle is fundamental to safety excellence. This is particularly essential as management layers are removed and self-managing concepts are introduced. A clear set of values provides the framework for responsible empowerment within guidelines.

The everyday stressing of safety values is one of the best ways for leaders to communicate their expectations, build safety awareness, and provide direction to the workforce. The beliefs must be lived and brought to life in the practices—"walk the talk"—or they are of little use.

The Results (Question 7)

Question 7 assessed the presence and influence of a set of safety values.

> **7.** Does your organization have well-established, written **safety values**—sometimes called beliefs and principles? "Written" means readily available in a document, posted on the bulletin board, etc. Check only one answer.
> ☐₁ Yes
> ☐₂ No
> ☐₃ I don't know

If you answered "No" or "I don't know," please go to question 8. If you answered "Yes," indicating that your organization does have written safety values, please check the statement below that best describes those values.

$\square_1$ We have safety values and they are up-to-date and well under-stood and have an important influence on safety.

$\square_2$ We have safety values and they have some influence on safety.

$\square_3$ We have safety values, but they are not used much and they have little influence on safety.

The respondents in the very safe companies had no doubt that their organizations have well-established safety values. At DuPont and Milliken, virtually 100% of the respondents answered "yes." The results of question 7 are shown in Fig. 6-12.

A specific set of safety values has been in place at DuPont for many years. The values are alive, and they are used regularly for discussion and communication. Similar sets of values for a dozen of the most important parameters (such as quality and integrity) are the foundation for the company's self-managing processes. Each employee at Milliken is given a refresher course on the company's safety values every year and is required to sign off, indicating understanding and acceptance.

At the unsafe companies, some people seemed unsure whether there were values or not. In the combined proportion of all respondents, none of the answers from the poorly performing companies was as high as the lowest for the very safe companies.

The answers to the second part of the question revealed that if there

The Presence of Safety Values			
	% Who Said Co. Has Safety Values		
	Yes	No	Don't Know
Best Result	99	1	0
Safe Co. Avg.	90	7	3
Unsafe Co. Avg.	50	31	19
Worst Result	38	53	9

Figure 6-12 The very safe companies have safety values. Those with poor safety are less likely to have them (Q7).

The Influence of Safety Values			
	% Who Said Safety Values Have ...		
	Important Influence	Some Influence	Little or No Impact
Best Result	88	10	2
Safe Co. Avg.	67	22	11
Unsafe Co. Avg.	12	28	60
Worst Result	3	26	71

Figure 6-13   Safety values have a strong influence in the safe companies but little in those with poor safety (Q7).[vii]

are values in the poorly performing companies, they are not of much influence (Fig. 6-13).

There was some spread in the results from the best-performing companies. Nevertheless, there is a clear separation between the results of the safe companies and those with poor safety. The lowest answer from the very safe companies was more than double the highest answer from the companies with poor safety. At DuPont and Milliken, where virtually all respondents said there were written values, almost all said they were influential. At Abitibi, safety values had been integrated with other values, and, at the time of the survey, they were just being articulated separately. This was one reason for the lower average for the very safe companies.

Comments on Question 7

Few companies use values effectively. Yet they are perhaps the best vehicle for clarifying intent and developing commitment. Values can be a powerful, practical tool for initiating and sustaining the course of change. The question was quite successful in discriminating between organizations with excellent and those with poor safety.

[vii] In computing the percentage of respondents who said that values were influential, those who said "no" or "don't know" to the first part of question 7 were assumed to believe the values were of little importance.

QUESTION 8: LINE MANAGEMENT RESPONSIBILITY — ACCOUNTABILITY FOR SAFETY

Safety is a line responsibility. Each executive, manager, or supervisor is responsible for and accountable for preventing all injuries in his or her jurisdiction, and each individual for his or her own safety and, in a less direct sense, for the safety of co-workers. (Belief 7)

This fundamental belief links the values of the company and the commitment of its leaders to the actions required by line management to implement them.

For management commitment to be translated into excellence, line management must take full responsibility for driving the safety agenda and be held accountable for injuries and incidents. "The line" or "line management" is the direct management line through which work is done — the leader, managers, supervisors *and individual workers.*[viii]

Line management must be committed to the vision and values and must see that they are understood and lived by in their jurisdictions. They must translate the objectives of the organization into specific objectives for themselves and their people. They are the main custodians of safety practices, seeing that they are meticulously implemented. They must seek out the best safety technology, hard and soft, for their unit's needs. They must take responsibility for ensuring that no one in their jurisdiction is injured. They must accept full accountability for an injury to any person in their area, not in a punitive way, but nonetheless clearly and unequivocally. The line must take responsibility for the effectiveness of auditing, injury investigation, return-to-work practices, recognition, safety rules and their enforcement and the other practices of safety management.

If line management is truly held accountable, safety will be included in the objectives, performance reviews, and pay of managers and supervisors and in the job descriptions and performance reviews of workers, too. Workers often perceive that their leaders are not really held accountable for injuries. To be credible, accountability must be visible.

[viii] This belief is mirrored in the "internal responsibility system" (IRS), which can be described as follows (the author's definition): "*The CEO is directly responsible and accountable for the safety of all employees. In a similar way, each manager and supervisor is directly responsible and accountable for the employees in his or her jurisdiction. Each employee is directly responsible and accountable for his or her own safety and in an indirect sense for the safety of co-workers. Each of these workplace parties is fully (100%) responsible and accountable but their responsibilities have different scope.*"

The internal responsibility system, with its cascading 100% responsibility, must be explained. Workers must know how their leaders are held accountable for results. For example, they should know, at least in a general way, how safety performance is reflected in their leaders' performance rating and pay. The accountability of workers then can be seen as a seamless part of the cascading responsibility of the line.

One consequence of line management taking responsibility is evident in what happens if an injury or incident occurs. Management does not find out about it from the medical department. Instead there is immediate follow-up by supervision, first to get the best care for the injured person, second to notify others who might be endangered by the same kind of incident, and finally to ensure return to work as quickly as possible. Medical support people, internal and external, understand and support this practice.

The full responsibility and accountability of line management for safety is not a paternalistic concept; it does not diminish the full responsibility of individuals for their own safety. It is an integral aspect of care and concern for the people in the organization.

Question 8 addressed this important concept.

The Results (Question 8)

8. Indicate the extent to which **line managers are held accountable** for injuries and safety incidents in their areas. (Line managers include such titles as supervisor, foreman, superintendent, team leader, etc., as well as manager.) Check only one answer.

In our organization:

☐$_1$ Line managers are held fully accountable for preventing injuries and incidents in their area. Safety performance has a direct effect on their performance rating and pay. This is a key part of our safety management.

☐$_2$ Line managers are held accountable for preventing injuries and incidents in their area but safety performance does not generally affect their performance rating and pay.

☐$_3$ Line managers are held accountable for injuries and safety incidents but only in a general way.

☐$_4$ While line managers take some responsibility for injuries and incidents in their areas, most injuries are attributed to individual error, bad luck, or unfortunate circumstances.

$\square_5$ Injuries and incidents are almost always blamed on individual error, bad luck, or unfortunate circumstances. Safety is much less important than business factors, such as costs and profits, in the assessment of managers' performance.

Even in the very safe companies, not all respondents felt that line management was held fully accountable for the safety of their people. Nevertheless, this critical question clearly separated the safe companies from those with poor safety (see Fig. 6-14).

In the companies with poor safety, only 12% of the respondents said that their management was held fully or fairly accountable and 39% felt that line management was not held accountable at all. In contrast, 75% of the respondents from the very safe companies reported that their management was held fully or fairly accountable. In Fig. 6-14, two "worst results" are cited, somewhat different but both very bad.

In most cases, workers perceived line management to have lower accountability than did management or supervision. Figure 6-15 shows the percentage of respondents who said that line management was held *fully* or *fairly* accountable.

The perception of workers was generally lower than that of managers and supervisors.

The results from one of the unsafe companies (Fig. 6-16) illustrate the radical difference in perception between workers and management that can exist when safety is very poor. Here the results for managers and supervisors were added together.

Managers and supervisors think that they are accountable. Workers do not. Many check off the answer that states that most injuries are

Line Management Accountability				
	% Who Said Line Management Is Held Accountable ...			
	Fully	**Fairly**	**Only Generally**	**Little Or Not At All**
Best Result	61	22	8	9
Safe Co. Avg.	44	31	11	14
Unsafe Co. Avg.	10	6	47	37
Worst Result 1	2	7	68	23
Worst Result 2	13	10	13	64

Figure 6-14 Line management is held more accountable for safety performance in the safe companies (Q8).

Line Management Accountability			
	% Who Said Line Management Is Held Fully Or Fairly Accountable ...		
	Managers	Supervisors	Workers	All
Best Result	100	94	68	83
Safe Co. Avg.	93	86	64	75
Unsafe Co. Avg.	23	22	8	16
Worst Result	14	6	0	4

Figure 6-15 Workers have a lower perception of management's accountability than do supervisors or managers (Q8).

Line Management Accountability Company with Poor Safety			
	% Who Said Line Management Is Held Accountable ...		
	Fully	Fairly	Only Generally	Little Or Not At All
Man & Sup	40	40	0	20
Workers	8	4	13	75

Figure 6-16 When safety is poor, workers don't think that management is held accountable at all (Q8).

attributed to individual error, bad luck, or unfortunate circumstances and that safety is much less important than business factors in the assessment of managers' performance. The responses of the workers are more believable than those of the managers and supervisors—the safety record of this company was very bad.

Comments on Question 8

This is one of the most important questions in the survey. It proved to be a good predictor of safety. When the responses from the very safe companies are compared with those of the companies with poor safety, there is a good correlation with safety performance. However, the responses at the very safe companies were not as good as the leaders at the companies expected. S&C management suggested that it might be a problem of terminology. They have stressed strongly that each person is fully responsible for his or her own safety. Unless the "internal responsibility system" is well understood, this might lead to the belief that supervision and management are not also responsible. Comments written in and raised in discussion groups indicate that this might be the case.

QUESTIONS 9 AND 10: INVOLVEMENT IN SAFETY ACTIVITIES AND EMPOWERMENT

Involvement of everyone in "doing things in safety" is the most powerful way to embed safety values and build safety awareness. (Belief 8)

This belief establishes the importance of involvement as one of the main vehicles for developing safety awareness and understanding of safety practices and for embedding commitment to safety values.

Involvement is a powerful concept, yet few companies take full advantage of it. Common sense might tell you that the best way to get things done is to assign the task to skilled people with special knowledge, safety specialists in this case. An *efficient* result may be obtained this way, but not the most *effective* result. Something as fundamental as safety cannot be a specialist function. When a team of supervisors and workers (with the stress on workers) develops a new safety practice, the wording may not be perfect and it may take more time to complete it, but it will be based on real work conditions. The workers involved will understand the practice, will feel a strong sense of ownership, and will be influential in persuading their peers of its benefits.

In very safe workplaces, everyone is involved, as a means of getting the safety work done and at the same time building understanding of safety and commitment to safety values. Narrowly based permanent committees are not allowed to take over. Workers and line supervision, not safety specialists, do most of the safety work. They are organized in committees and task forces, and their membership rotates. This concept is not unique to safety; it is the essence of self-management.

Involvement is not just consultation or passive participation. There is a great difference among the following paths:

1. The safety specialist or the operating supervisor develops the practice.

2. The specialist or the supervisor asks the opinion of workers, then develops the practice.
3. The specialist or the supervisor adds workers to a committee that develops the practice.
4. A cross-sectional team comprised largely of workers is given the mandate to develop the practice, within the boundaries that it must satisfy both safety and job effectiveness. The supervisor and the safety specialist participate on the team and provide guidelines and advice, but the workers do most of the work.

Although the second and third alternatives may seem like involvement to managers and supervisors, they do not seem so to workers. Likely the work would be largely done by staff people, who are more used to developing procedures. Only the last alternative is true involvement, hands-on "doing the task" and taking pride in and ownership of the result.

Companies sometimes cite the difficulty of getting workers involved and, in some cases, their refusal to become involved. In the safest companies, this is not a real problem because workers see the benefit to themselves and to their peers of working purposefully on safety. One way or another almost everyone becomes involved.

Many jurisdictions, including most of the Canadian provinces, have legislated workplace joint health and safety committees. The legislation usually requires equal numbers of management and of workers, the latter selected by the union or elected by workers if there is no union. Governments and others cite these committees as important vehicles for involvement. However, this involvement is far narrower and more limited in scope than is envisaged here and practised by the safest companies.

Involvement and empowerment go hand in hand. It is difficult to conceive of true involvement not leading to empowerment or of empowerment without involvement. As companies move increasingly to self-managing systems to capitalize on the great advantages in motivation and productivity, safety, like everything else, must be managed differently. The safety values provide a principled framework. Through involvement and training, employees are given knowledge and skill in safety. They are empowered to self-manage many aspects of safety, as individuals and as teams.

A number of questions addressed the subject of involvement. Question 9 dealt directly with involvement in two different ways. Question 10 asked about empowerment. Then, in question 16 about audits, involvement was raised again.

The Results (Question 9, Part 1): Involvement in Safety Activities

The first part of this question asked directly about involvement.

9. How actively have you been **involved** in safety activities **in the last year**? Involvement means not just attending meetings but participation in doing things in safety such as being on a committee, participating in an investigation, or helping put together safety rules. Check only one answer.

   $\square_1$ Deeply involved

   $\square_2$ Quite involved

   $\square_3$ Moderately involved

   $\square_4$ Not much involved

   $\square_5$ Not involved at all

In all of the companies surveyed, workers reported lower involvement than did managers and supervisors. Figure 6-17 illustrates this finding.

The extent of involvement was much greater in the very safe companies than in the poorly performing companies, as is shown in Fig. 6-18.

In the companies with poor safety, involvement of workers was very low, virtually nonexistent (Fig. 6-19).

It is from involvement that these organizations stand to gain the most. Involvement is the best way for people to learn and absorb the

Involvement in Safety Activities			
	% Who Said They Are Deeply or Quite Involved in Safety Activities		
	Managers	**Supervisors**	**Workers**	**All**
Best Result	89	80	68	74
Safe Co. Avg.	75	69	42	56
Unsafe Co. Avg.	54	31	19	29
Worst Result	17	21	16	22

Figure 6-17   Workers are less involved in safety activities than are supervisors or managers (Q9).

Involvement in Safety Activities			
	% Who Said They Are Involved ...			
	Deeply	Quite	Moderately	Not Much Or Not At All
Best Result	46	28	18	8
Safe Co. Avg.	28	28	20	24
Unsafe Co. Avg.	8	21	21	50
Worst Result	3	13	7	77

Figure 6-18   There is a radical difference in involvement between the safe companies and those with poor safety (Q9).

Involvement of *Workers* in Safety Activities			
	% Of *Workers* Who Said They Are Involved ...			
	Deeply	Quite	Moderately	Not Much Or Not At All
Best Result	39	29	20	12
Safe Co. Avg.	22	20	23	35
Unsafe Co. Avg.	5	14	16	65
Worst Result	4	8	0	88

Figure 6-19   In companies with poor safety, workers are not much involved in safety activities (Q9).

values of the organization. These companies are not taking advantage of this powerful way to build value for safety.

The Results (Question 9, Part 2): Involvement in *Specific* Safety Activities

Involvement is subjective. As people become more involved, their standards become higher. Thus the involvement in the safest companies may be higher than was reflected in the answers. In the poorly performing companies, the results were probably biased in the other direction. Because of this, a second part to question 9 asked more specifically whether respondents had been on a standing committee, task force, or team on safety in the past 2 years.

In the last **two years**, have you been on either a standing safety committee (for example, a JH&SC, a rules and procedures committee, a safe driving committee, etc.) *or* a specific task force or team (for example, a team formed to review the safety rules of an area)?

☐₁ Yes

☐₂ No

This question was added after the research had been partly completed, and only one company with very poor safety was polled. However, the responses gave the same message: workers are involved in safe companies and not in unsafe companies (Fig. 6-20).

The goal of the company with the best result (Milliken) is 100%—every employee involved all the time on a meaningful assignment in safety. They define involvement as work beyond the normal safety of the individuals' jobs, for example, participation on a committee, task force, or special team or taking assigned responsibility for auditing.[ix] Milliken measures involvement regularly, and the company claimed to have reached 70% in early 1998. That number tallied well with the 68% of workers (in Fig. 6-19) who said that they were deeply or quite involved and the 71% of workers that reported (in Fig. 6-20) that they had been involved in specific activities in the last 2 years. (Although the close correspondence is coincidental, this is gratifying corroboration of the questionnaire technique.) There are no safety specialists in the company's several dozen plants: the safety work is done by teams of managers and workers. In another of the very safe companies,

Involvement of *Workers* in Specific Safety Activities		
	% Of *Workers* Involved in Specific Safety Activities in Last Two Years	
	Yes	No
Best Result	71	29
Safe Co. Avg.	39	61
Unsafe Co. Avg.	8	92
Worst Result	8	92

Figure 6-20 The striking contrast in involvement of workers in safe companies and workers in companies with poor safety (Q9).

[ix] In both Milliken and DuPont, typical plants have 6–10 standing committees with such responsibilities as process hazards, off-the-job safety, and rules and procedures.

Involvement of *Workers* in Safety Audits			
	% of *Workers* Who Said That They Are ...		
	Regularly Involved	**Have Some Involvement**	**Not Involved At All**
Best Result	62	34	4
Safe Co. Avg.	35	36	29
Unsafe Co. Avg.	0	8	92
Worst Result	0	8	92

Figure 6-21 Workers are quite involved in safety audits in the safe companies, not much in companies with poor safety (Q16).

the goal is 50% involvement. Both have LWIF below 0.1 and TRIF below 1.

Question 16 on workplace inspections/audits also asked about involvement (Fig. 6-21).

In the company with the best result, almost everyone, and over 90% of workers, reported that they were involved in organized audits of their workplace. In that company, audits include assessment of safety systems and of safety awareness of people. This excellent opportunity to involve people in safety is not being used at the poorly performing companies. (This question was asked of only one company with poor safety.)

Even among the very safe companies, some do not fully utilize involvement, the most effective way to build the safety culture into everyone's mind-set. Some feel that having workers do things that could perhaps be done better and more quickly by specialists is wasteful. Quite the reverse is true: Involvement gets the safety work done while building understanding and commitment into everyone's mind.

Question 10: The Extent of Empowerment

A relatively full explanation of the meaning of empowerment in the context of safety was included in question 10. It defines the scope of the empowerment that is envisaged.

10. To what extent are you **empowered to take action** to ensure your own safety and that of others with whom you work? *Empowered* means that you are expected to take whatever action is required to avoid injuries to yourself or to others, including shutting down equipment. It means that you are empowered to fix unsafe situations within the scope of your job, expected to take whatever action is required in urgent cases, and encouraged (expected) to make recommendations even if the situation does not relate to your job. It means that your organization has created an environment in which everyone is encouraged to contribute to improvement. It means that you have had safety training and involvement in safety activities that give you confidence to act.

On the other hand, *not empowered* means that for safety matters, you are expected to stick pretty strictly to the confines of your specific job description. Check only one answer.

☐₁ Fully empowered
☐₂ Quite empowered
☐₃ Moderately empowered
☐₄ Not very empowered
☐₅ Not at all empowered

The results showed a major difference between the safe and the poorly performing companies (Fig. 6-22).

Question 10 was developed after most of the questionnaire surveys on the companies with poor safety had been completed, so Fig. 6-22 refers to just one unsafe company. Nevertheless the contrast is striking.

Empowerment of *Workers*				
	% Of *Workers* Who Said They Are Empowered ...			
	Fully	Quite	Moder-ately	Not Very Not At All
Best Result	81	13	3	3
Safe Co. Avg.	67	20	12	1
Unsafe Co. Avg.	17	0	25	58
Worst Result	17	0	25	58

Figure 6-22 Workers in safe companies are more empowered to affect safety than in companies with poor safety (Q10).

Comments on Questions 9 and 10

Involvement is the most powerful way to build commitment to safety and to support a step change in performance. Yet companies with poor safety and even those trying to improve their performance do not understand it or utilize it well. Thus it was important to measure involvement as quantitatively as possible and to do it in several ways. It was thus gratifying to find that the questions on involvement and empowerment discriminated very well between the safe and the unsafe companies.

QUESTION 11: SAFETY TRAINING

Safety training is an essential element in developing excellence. It complements but cannot replace "learning by doing," in itself a method of training. (Belief 9)

This belief defines the necessity for thorough training of the workforce in safety while stressing that training is not a substitute for involvement but complementary to it.

Training is one of the cornerstones of an effective safety culture. In the safest companies, everyone is trained regularly and thoroughly in specific job safety techniques and in more general safety practices. Development of safety awareness and embedding of safety attitudes are built into all of the formal training processes. These companies understand that the best training involves a "learn—do—learn" cycle so that knowledge learned is quickly consolidated by practical use. They also know that the best training comes from involvement in "doing things in safety," so training is integrated with involvement in safety activities. In this sense, training is not in itself a panacea; it is part of an integrated approach to safety. Often training is seized on as a solution to poor safety performance—the "magic bullet." However, introducing extensive training without integrating what is learned directly and immediately into the workplace will not solve the problem.

In the very safe companies, new employees are given extensive safety training, immediately after hiring, including discussion of the organization's safety values as well as specific job safety training. As a result of this conditioning, the incidence of injuries among new workers is as low as that of the rest of the workforce.

The Results (Question 11)

11. Indicate the extent to which you have received **training** in safety and occupational health **in the last two years**. Training includes formal training courses away from the job and organized on-the-job training. Check only one answer.

$\square_1$ Thorough and extensive training

$\square_2$ Considerable training

$\square_3$ Some training

$\square_4$ Little training

$\square_5$ No training

Figure 6-23 shows that training is much more intensive at the very safe companies than in those with poor safety.

Among the very safe companies, respondents in all job categories reported a substantial level of training, with workers generally reporting a somewhat *higher* level than supervisors or managers. Not only was training low in all categories among the poorly performing companies, but workers generally reported somewhat *lower* levels of training than managers. Thus the results for workers show an even greater spread between the results of the very safe and the poorly performing companies (Fig. 6-24).

Milliken respondents reported the "best result"—the highest level—for worker training. The company requires each of its several thousand employees to have at least 32 hours of formal training in safety each year. This level of training is considerably higher than that of any of the other very safe companies. The Fort Frances Mill of Abitibi-

Safety Training					
	% Who Said Their Training in Last Two Years Has Been ...				
	Exten-sive	Consid-erable	Some	Little	None
Best Result	61	31	5	2	1
Safe Co. Avg.	19	38	28	12	3
Unsafe Co. Avg.	3	9	36	25	27
Worst Result	3	7	13	27	50

Figure 6-23 The safe companies do much more safety training than do companies with poor safety (Q11).

Safety Training of *Workers*					
	% Of *Workers* Who Said Their Training in Last Two Years Has Been ...				
	Exten-sive	Consid-erable	Some	Little	None
Best Result	71	23	4	1	1
Safe Co. Avg.	26	43	24	6	1
Unsafe Co. Avg.	1	7	29	25	38
Worst Result	0	4	8	29	59

Figure 6-24   The striking contrast in training for workers (Q11).

Consolidated, one of the other companies in the benchmark group, requires each employee, from manager to worker, to take a one-day refresher course in safety every year. This training takes place in groups that include supervisors, managers, and workers. It is planned and conducted largely by workers on rotating special assignment. This is in addition to normal safety training. DuPont Canada gives summer relief students several days of training before they begin work. Their safety record is as good as that of the regular population.

Comments on Question 11

The answers discriminate very well between the very safe companies and those with poor safety, thus helping to validate the question and the model. Very safe companies train their people; companies with poor safety do not.

QUESTION 12: THE FREQUENCY AND QUALITY OF SAFETY MEETINGS

Regular, effective safety meetings involving all people in the workplace are an essential part of good safety. (Belief 11)

One of the most important practices in developing safety awareness is the safety meeting system. Safety meetings provide a forum for reinforcing the beliefs and practices, for exposing and discussing safety issues, and for providing new information on safety techniques. They should be held frequently and regularly. Attendance should be a job requirement for everyone—clerical, administrative, and management people as well as workers. Safety meetings should be open, participative, and stimulating, a standard not usually met. Often workers judge their safety meetings to be routine, boring, and not very useful. It is supervision's responsibility to see that safety meetings are held regularly and are fully attended and that they provide a real contribution to safety. In the very safe companies, workers usually plan and conduct most safety meetings. They are naturally motivated not to bore their peers.

The Results (Question 12)

There were three parts to the question on safety meetings dealing with frequency, attendance, and quality. This question was added to the questionnaire after some of the research had been completed, so only three of the very safe companies and one of the companies with poor safety were polled.

12. Are **safety meetings** held regularly in your workplace? If so, how often?
    $\square_1$ Every week or every two weeks
    $\square_2$ Every month
    $\square_3$ Every two months
    $\square_4$ Less frequently than every two months
    $\square_5$ Not regularly held

    Do you attend the safety meetings regularly?
    $\square_1$ Yes
    $\square_2$ No

    How do you rate the **quality and effectiveness** of the safety meetings? Consider how well attended they are. Consider the content of the meetings and the extent of involvement of people in developing and conducting them.
    $\square_1$ Excellent
    $\square_2$ Good
    $\square_3$ Satisfactory
    $\square_4$ Poor
    $\square_5$ Very poor

In Figs. 6-25, 6-26, and 6-27, the responses from *workers* are discussed. The results for all respondents were not that different from the results for workers.

In the very safe companies, safety meetings for workers are held at least once a month and usually more often. The standard is every week or every 2 weeks. This was confirmed in the survey (Fig. 6-25). In contrast, two-thirds of the workers in the one poorly performing company surveyed said that they did not have regular safety meetings at all. A comment from a worker in that company: "I haven't seen one. I have never been to a safety meeting nor have I ever heard of one being held."

The same striking difference was observed in the question referring to attendance (Fig. 6-26).

The average for the very safe companies was pulled down by a relatively low result from S&C Electric, where safety meetings are held less formally in the workplace.

The perception of the quality of safety meetings is given in Fig. 6-27.

In the one company with poor safety polled with this question, if safety meetings were held at all they were not perceived by workers as very useful. In the very safe companies, more than 80% of workers perceived safety meetings to be excellent or good. In these companies, workers are usually involved in planning and conducting most safety meetings. The standard is high participation and stimulating, useful meetings.

Workers' Views on the Frequency of Safety Meetings					
	% Of *Workers* Who Said Frequency of Safety Meetings Is ...				
	1 or 2 Weeks	Monthly	Bi-monthly	<Than Bi-monthly	Not Regular
Best Result	99	0	0	1	0
Safe Co. Avg.	57	31	4	6	2
Unsafe Co. Avg.	5	24	5	0	66
Worst Result	5	24	5	0	66

Figure 6-25 Safety meetings are held more frequently in the very safe companies (Q12).

Attendance of *Workers* at Safety Meetings		
	% Of *Workers* Who Said They Attend Safety Meetings Regularly	
	Yes	No
Best Result	96	45
Safe Co. Avg.	74	26
Unsafe Co. Avg.	10	90
Worst Result.	10	90

Figure 6-26   Attendance at safety meetings is high in the very safe companies, low in those with poor safety (Q12).

Workers' Views on the Quality of Safety Meetings			
	% Of *Workers* Who Said Safety Meetings Are ...		
	Excellent	Good	Satisfactory	Poor or Very Poor
Best Result	40	48	12	0
Safe Co. Avg.	35	48	17	0
Unsafe Co. Avg.	6	11	22	61
Worst Result	6	11	22	61

Figure 6-27   Workers considered the quality and effectiveness of safety meetings to be high in the very safe companies (Q12).

Comments on Question 12

Because only one company with poor safety was polled, the question has not been well validated. However, the limited results were as expected: a striking difference between the results from the very safe companies and the one result from a company with poor safety. This difference would almost certainly be confirmed by a larger sample. This practice is so well entrenched as an integral part of the safety culture in the very safe companies that it is hard to imagine excellent safety without effective safety meetings.

QUESTION 13: SAFETY RULES

Comprehensive, up-to-date safety rules, crafted with broad participation and consistently applied, are essential for excellence in safety and also assist in doing the job well. (Belief 12)

In the ideal case, safety rules and practices are carefully constructed with full participation of the workforce, regularly updated, and thoroughly communicated. Everyone is expected to follow them. The rules are reinforced by disciplinary action geared to the infraction and to the circumstances. The safest companies approach this standard.

Rules should be separated from *practices*. Rules are the way that tasks *must* be done, every time. There is no tolerance for taking a short cut with lock, tag, and try rules in an industrial plant any more than there is for failing to stop at the stoplight in front of a school. Rules should be kept to the necessary minimum, to those essential safety requirements that must be followed and will be enforced.

Practices describe the *recommended* way to do things: "It is a good practice to check the condition of your tools before you begin work." With this separation of rules and practices, rules can be rigorously enforced. There will be more practices than rules, and some practices will be optional, at the judgement of supervision or workers. If some practices are compulsory, that must be clearly spelled out.

Question 13 dealt with the quality of the safety rules and their observance.

The Results (Question 13, Part 1): The Quality of Safety Rules

The first part of the question dealt with the quality of safety rules.

13. Please consider the **quality of the safety rules** in your organization. High-quality rules are up-to-date and clearly written. They are well understood by those doing the work and help them to do the job well and safely. Check only one answer.

The quality of our safety rules is:

☐₁ Excellent
☐₂ Good
☐₃ Satisfactory
☐₄ Poor
☐₅ Very poor

Workers' Views on the Quality of Safety Rules				
	% of *Workers* who Said Quality of Rules Is ...			
	Excellent	Good	Satis-factory	Poor Or Very Poor
Best Result	67	30	3	0
Safe Co. Avg.	53	43	4	0
Unsafe Co. Avg.	12	29	33	26
Worst Result	9	18	27	46

Figure 6-28   Safety rules are seen to be of good quality in safe companies, of much lower quality when safety is poor (Q13).

In the figures that follow, the results shown are the perceptions of workers. In the very safe companies, there was little difference among the answers of managers, supervisors, and workers, although workers had a somewhat more positive opinion. In the companies with poor safety, managers and supervisors consistently had a better view than workers of the quality of the rules and of their observance.

The difference between the results from the very safe companies and those with poor safety is shown in Fig. 6-28.

Almost all the respondents from the very safe companies said that their safety rules were good or excellent. Only 40% of the respondents in the companies with poor safety considered their rules to be at that high level, and one-quarter felt that they were poor or very poor. This is a very radical difference. Safety rules are intended to protect people from injury, but how can they if they are of poor quality?

In the ideal case, workers respect the rules because they understand them and know that they are crafted to protect them from injuries. Workers will have participated in making and revising the rules. If workers feel that the rules are out of date or of poor quality, they very likely are—and not only will they provide less protection but also people will be less inclined to follow them.

The Results (Question 13, Part 2): The Extent to Which Safety Rules Are Obeyed

To what extent are the **safety rules** of your organization **obeyed**? Check only one answer.

$\square_1$ All safety rules are obeyed without exception.
$\square_2$ People generally obey the safety rules.
$\square_3$ The safety rules are guidelines, sometimes followed, sometimes not.
$\square_4$ The safety rules are often not obeyed.
$\square_5$ People pay little attention to the safety rules.

The "without exception" phrase was used intentionally, knowing that it would be questioned, as it often has been when this questionnaire has been used in consulting. The comments usually go along the line of: "Be reasonable. There are always exceptions. There must be some room for individual decision." Yet who would want to give drivers passing their children's school the option of deciding that they wouldn't stop at the stoplight because it was after regular hours or for whatever other reason?

As would be expected, workers in the companies with poor safety described themselves as less likely to observe their safety rules than did workers in the very safe companies (Fig. 6-29).

In companies with poor safety, management likely sets a poor example. A worker from one of the companies polled commented, "I think that a lot of improvements could be made, starting with the committee and the supervisors. They stress that the workers are being safe with glasses and hard hats, when often they themselves don't follow the safety standards. I think that they should set the example for the workers. If they don't have to follow the safety standards, then why should the workers?" In the safe companies, almost all the workers

Workers' Views on the Observance Of Safety Rules				
	% Of Workers Who Said That Rules Are Followed ...			
	Without Exception	Generally	Sometimes	Often Not, Little Attention
Best Result	37	60	0	3
Safe Co. Avg.	25	64	10	1
Unsafe Co. Avg.	4	35	54	7
Worst Result	0	27	64	9

Figure 6-29   Safety rules are not well observed in companies with poor safety (Q13).

(about 90%) said that the rules are generally followed. One-quarter were prepared to check off "without exception."

Comments on Question 13

For both parts of this very important question, the results correlated well with the safety of the responding organizations.

QUESTION 14: ENFORCEMENT OF SAFETY RULES

Disciplinary action for safety infractions is an essential part of good safety. Its intent is not punishment or retribution but the correction of unsafe behaviour, the demonstration of the standards of the organization and the weeding out of those who will not accept their responsibility for safety excellence. (Belief 13)

If the safety rules are to have meaning, they must be enforced. The converse is also true. Unless the rules are clear, well thought-out, thoroughly communicated, and understood and accepted by the workforce, attempting to enforce them with disciplinary action, will be a mistake, an unfair one.

Disciplinary action is an important element of excellence in safety. It is defined as "a range of actions, from a cautionary conversation through to more severe action such as termination." Discussion with the rule-breaker is by far the most frequent type of action. Many would not consider such discussion to be disciplinary action, but it is.

In the safest companies, the goal is consistent disciplinary action for *all* infractions of rules, although even they do not reach this ideal standard. Some companies have a "zero tolerance" (mandatory termination) policy for breaking important safety rules. Of course, when safety is excellent and the rules are of good quality, understood, and respected, and when the consequences of breaking them are thoroughly communicated, disciplinary action is seldom required.

Should there be disciplinary action when the rule-breaker sustains an injury? Many would say the injury itself is enough punishment. But punishment is not the goal. The intent is to reinforce the concept that rules must not be broken. In some instances, disciplinary action is appropriate even though an injury has occurred. The transgression may be serious enough to result in civil or criminal action. Should a conscientious company take it less seriously? There are positive approaches to disciplinary action, such as requiring the rule-breaker to conduct a safety meeting on the issue involved.

Disciplinary action for infractions of safety rules is a controversial subject. Unfortunately, that means that managers and supervisors often avoid bringing it out into the open. Workers fear that it will be arbitrary and applied only to them. Without good safety rules and well-thought-out principles of enforcement, this is often true.

There is a legitimate fear that disciplinary action will drive incidents and injuries underground. Yet deliberate breaking of safety rules must not be ignored. The answer is to address the issue forthrightly and openly. Good employees, intent on working safely, have nothing to fear. No one wants to have people around who willfully break safety rules, endangering themselves and others. The best approach is to create the half-dozen beliefs on which disciplinary action will be based, communicate them thoroughly, and apply them fairly for *all* infractions.[x]

The Results (Question 14)

14. Indicate your opinion of the way that **disciplinary action** is used in your organization for infractions of safety rules or practices. "Infraction" means breaking a safety rule or not following a standard practice. Disciplinary action refers to the range of actions, from a cautionary conversation or warning through to more severe action such as termination. Check the one answer you believe is the most accurate.

   $\square_1$ Disciplinary action, related to the seriousness of the infraction, is taken for all safety infractions.

   $\square_2$ Disciplinary action is taken only for serious safety infractions.

   $\square_3$ Disciplinary action for safety infractions is applied arbitrarily and inconsistently.

   $\square_4$ Disciplinary action is seldom taken for safety infractions.

The results in Fig. 6-30 indicate that the safest companies approach the standard of enforcement of all rules with some type of disciplinary action.

---

[x] An example of one of the beliefs specific to disciplinary action is the following: "Disciplinary action applies to all levels of the organization. Management and supervision are accountable for the creation of a safe, orderly environment. Failure to do so will result in disciplinary action."

Workers' Views on the Enforcement of Safety Rules			
	% Of *Workers* Who Said Disciplinary Action Taken For ...		
	All Infractions	Serious Infractions	Not Consistent Arbitrary	Seldom Taken
Best Result	70	17	13	0
Safe Co. Avg.	53	22	10	15
Unsafe Co. Avg.	18	18	20	44
Worst Result	4	12	4	80

Figure 6-30   Enforcement of safety rules is much more consistent in companies with good safety than in those with poor safety (Q14).

Milliken has a "zero tolerance" (mandatory termination) policy for breaking any of its most important safety rules, such as the "lock, tag, and try" procedure designed to safeguard people working on equipment. DuPont has also applied a forthright approach to this sensitive issue. It has crafted a set of beliefs and principles about disciplinary action and communicated them widely. It aims to use disciplinary action for all infractions.

Comments on Question 14

Disciplinary action is so important that it would have been disappointing if the results had not correlated with safety performance.

QUESTION 15: INJURY AND INCIDENT INVESTIGATION

Every injury or incident is an opportunity to learn and improve. Thorough, participative processes of investigation are a cornerstone of safety. (Belief 14)

This belief is written as a mirror image of the belief that is one of the foundations of quality management. Thus the word *every* is used intentionally to mean 100%. If there is leeway to waive an investigation, some of the more embarrassing incidents, from which a great deal can be learned, might be missed. The investigation can be in less depth if the incident is a repeat. Even then, the question of why the incident recurred must be asked.

The careful tracking of incidents and injuries provides an indication of trends in the safety awareness of the workforce. The safest companies meticulously investigate and record major and minor injuries and incidents with potential for injury. They keep track of the frequency of all injuries and incidents, including first aid cases. The frequency of incidents that do not result in injury provides a *leading* indicator.

Injury investigation provides an opportunity for involvement. The manager or supervisor responsible for the area, not a safety specialist, should lead the investigation. Unless there are sensitive personal issues, other people in rotation, such as other managers, staff, and workers should be included. If possible, the injured person and other workers should be part of the team (while keeping it to manageable size). Involvement adds perspective and provides another opportunity for the participants to "learn while doing."

The purpose of the investigation is to learn what really happened, not to lay blame. However, the team should be forthright in its diagnosis, or the lessons will not be learned. If the cause of the incident was individual carelessness, that designation should not be avoided.

The safest companies are very thorough in communicating the results of the investigations, defining the action to be taken (or not taken) and taking quick action.

The Results (Question 15)

15. To what extent are injuries, safety incidents, near misses, and the like in your organization **investigated** and reported and action taken? Check only one answer.

   $\square_1$ All injuries and incidents are thoroughly investigated and the recommendations are implemented.

   $\square_2$ Most injuries and incidents are investigated and most of the recommendations are implemented.

   $\square_3$ Many of the injuries and incidents are investigated and some of the recommendations are implemented.

   $\square_4$ Only the most serious injuries and incidents are investigated.

   $\square_5$ Injuries and incidents are not usually investigated.

The results in Fig. 6-31 show a radical difference in the thoroughness of investigation between the very safe companies and those with poor safety records. In the very safe companies, a high proportion of the

Investigation of Injuries & Incidents					
	% Who Said Injuries & Incidents Investigated ...				
	All	Almost All	Many	Only Serious	Usually Not Done
Best Result	85	12	1	2	0
Safe Co. Avg.	68	27	4	1	0
Unsafe Co. Avg.	17	35	25	18	5
Worst Result	17	20	20	30	13

Figure 6-31   Injuries are more thoroughly investigated in the safe companies than in those with poor safety (Q15).

Investigation of Injuries & Incidents				
	% Who Said *All* Injuries & Incidents Are Investigated			
	Managers	Supervisors	Workers	All
Best Result	77	80	92	85
Safe Co. Avg.	70	58	70	68
Unsafe Co. Avg.	15	26	16	17
Worst Result	0	11	9	11

Figure 6-32   Workers and management generally agreed on the level of investigation of injuries and incidents (Q15).

respondents said that *all* injuries were investigated and virtually everyone said that either *all* or *almost all* were investigated. The unsafe companies are losing one of the best opportunities to learn how to avoid injuries.

In the answers to this question, there were no consistent differences between the opinions of management (managers and supervisors) and the opinions of workers, as shown in Fig. 6-32.[xi]

The best results for this question came from Milliken. Not by chance. Their thorough investigation process calls for seeking answers to What,

[xi] Note that the worst result in Fig. 6-32 is for a different company than the worst result in Fig. 6-31.

Where, When, Who, How, and *Five Levels of Why*! First, *why* did it happen? Because the worker didn't follow the safety procedure precisely. Second, *why* didn't the worker follow the safety procedure? The process continues, forcing the investigation team to peel back enough layers to get to the real causes. The usual result of this level of search is a broadening of the causes to include more than the individual involved.

Comments on Question 15

The results correlated well with the safety performance of the companies.

A legitimate criticism of this question is that it combines three things—investigation, reporting, and action to correct the defects. The same point could be made about other questions. They were designed this way to avoid an overly long, complicated questionnaire. Also, it would be expected (and it is the author's experience) that organizations that investigate thoroughly would communicate the results and follow up on them.

QUESTION 16: WORKPLACE AUDITS/INSPECTIONS

Auditing the workplace for physical defects, for the effectiveness of safety systems, and for the safety awareness of the people who work there is a valuable safety practice. (Belief 15)

Like safety meetings, workplace audits-inspections can be a very useful element in developing safety excellence. Yet they seldom are. Often they are done by supervision or by safety specialists, and usually they concentrate on the physical aspects of the workplace. This is useful, but it is perhaps the least of the potential benefits, and often, if audits are restricted to physical things, they become routine and boring. They can be brought to life and made a vital contributor to safety by:

- Including auditing of the safety awareness of the workforce
- Broadening audits to include assessment of the presence and effectiveness of the safety systems and practices
- Involving the whole workforce in regular rotation in conducting the audits
- Reporting widely the results and the lessons learned

If we believe that "95% of safety is in the head," in the attitudes of people and in their behaviour, audit systems should reflect that belief. A study of several hundred injuries found that only 3% could have been prevented by a prior audit of the conditions in the workplace (27). The safest companies stress assessment of the effectiveness of safety systems and of the safety awareness of individuals as well as the state of equipment. In some cases, audit results are "quantified" to form a leading indicator of trends (points assigned according to the severity of the observed situation). They also provide an opportunity to build knowledge and awareness through involvement.

The word "audit" is used intentionally, particularly if it involves teams in rotation. Physical safety defects and problems with safety systems that present an immediate hazard should be brought to the attention of the person responsible for the area. If unsafe behaviour is observed, the individual should be told about it on the spot. However, there must be no identification of the person involved. If a generic note is made in the report to alert the rest of the workplace parties to a problem, it should leave no clue as to the individual's identity. The purpose of the process is not to supervise. Workers, quite rightly, will not participate unless confidentiality is scrupulously observed.

Even in safe companies, involvement in safety audits-inspections is not as fully used as it might be. In the safest companies, almost everyone is involved in regular, organized audits-inspections of their workplaces.

The Results (Question 16)

Question 16 addresses two aspects of audits-inspections, their quality and effectiveness and the involvement of people in them.

16. Indicate the extent that you are **personally involved in safety audits** and inspections of the workplace. Involvement means participation on a regularly scheduled, organized basis, not informal, personal inspections. Check only one answer.
   $\Box_1$ Regularly involved
   $\Box_2$ Some involvement
   $\Box_3$ Not involved at all

How do you rate the **quality and effectiveness** of the safety audit and inspection system? Consider the frequency and thoroughness of the inspections, the extent of participation of the workforce, and the extent to which safety behaviour is observed, as well as physical conditions, the thoroughness of the follow-up, and the overall effectiveness in helping develop a safer workplace. Check only one answer.

$\square_1$ Excellent

$\square_2$ Good

$\square_3$ Satisfactory

$\square_4$ Poor

$\square_5$ Very poor

Involvement in safety audits-inspections is reported in Fig. 6-33. This question was added after much of the data had been collected for the companies with poor safety. Thus all five of the very safe companies were polled but only one company with poor safety.

Even in the very safe companies, involvement in safety audits is not as fully developed as it might be. In DuPont and Milliken, the two companies with the best results, almost everyone (over 90% of workers) said that they were involved regularly or at least had some involvement in organized audits of their workplace. In these companies, audits have been extended to include assessment of safety systems and of safety awareness of people. In the one company with poor safety that was polled, involvement was very low.

The results on the involvement of *workers* is given in Fig. 6-34. In the company with poor safety, workers said that they were not much involved. This company is missing a good opportunity to help

Involvement in Safety Audits			
	% Who Said That They Are ...		
	Regularly Involved	Have Some Involvement	Not Involved At All
Best Result	68	27	5
Safe Co. Avg.	40	31	29
Unsafe Co Avg.	7	10	83
Worst Result	7	10	83

Figure 6-33 In the very safe companies, safety audits are an important vehicle for involvement (Q16).

Involvement of *Workers* in Safety Audits			
	% of *Workers* Who Said That They Are ...		
	Regularly Involved	Have Some Involvement	Not Involved At All
Best Result	62	34	4
Safe Co. Avg.	35	36	29
Unsafe Co Avg.	0	8	92
Worst Result	0	8	92

Figure 6-34   In the safe companies, workers are involved in safety audits of their workplaces (Q16).

Quality of Safety Audits					
	% Who Said Quality Of Audits-Inspections Is ...				
	Excellent	Good	Satisfactory	Poor	Very Poor
Best Result	43	48	9	0	0
Safe Co. Avg.	29	51	18	1	1
Unsafe Co. Avg.	3	14	31	38	14
Worst Result	3	14	31	38	14

Figure 6-35   The quality of safety audits was deemed to be good in the safe companies (Q16).

people learn safety while they fulfill a useful function in seeking out hazards.

The perception of the quality of safety audits is reported in Fig. 6-35.

The quality was judged to be good or excellent in the safe companies and generally poor in the one unsafe company that was polled.

Comments on Question 16

The question gave results in line with safety performance. More data for companies with poor safety are needed to provide better confirmation.

QUESTION 17: MODIFIED DUTY AND
RETURN-TO-WORK SYSTEMS

Excellence in safety is enhanced by thorough efforts to find modified duties for injured people who cannot do their regular jobs but who can safely do other work and by comprehensive initiatives to assist in rehabilitation and ensure early return to work. (Belief 16)

In many workplaces, the modified duty and return-to-work systems are as contentious an issue as disciplinary action. Yet these systems can and should be valuable contributors to excellence in safety. If well managed, they operate to everyone's advantage. Workers recover more quickly if they continue to work *at tasks appropriate to their injury or disability*. Companies avoid the loss of valuable skills, and the cost is lower for the public agencies such as Workers' Compensation Boards.

In the earlier discussion of the role of line management, the responsibility of the supervisor or stand-in if an injury or incident occurs was outlined. He or she must be on the scene as soon as possible, preferably immediately, to:

1. See that the injured person gets the best care,
2. Assess whether the injury has implications for other people or other workplaces
3. *Assess whether the person can return to work and, if not, what steps are required for proper treatment and early return, either to the regular job or to work modified to suit the situation (in consultation with medical personnel)*

The conscientious application of the principle of early return to work helps build the understanding that injuries should not happen and that it is line management's responsibility to see that they do not. If they do happen, line management, not safety or medical specialists, is accountable and must take charge of remediation.

The safest companies reject the idea of an automatic entitlement to time off for an injury. There must always be a careful assessment of whether modified duties or return to work are medically advisable, but there should be no "entitlement" to time off. What workers are entitled to is a safe workplace in which injuries do not happen and appropriate measures are taken in the few instances when they do. Management's "feet should be held to the fire" to see that they deliver these real entitlements.

With these qualifications, early return to work, in modified duties if necessary, is usually best for the worker and the organization. These systems seldom receive full approval of people in the workplace, yet they should not be as unpopular with workers as they are. If they are well managed, thoroughly explained, and operated fairly, they should be accepted by workers as being in their long-term interest.

The Results (Question 17)

17. Good management of workplace safety includes strong efforts to find temporary **modified duties** for injured people who cannot do their regular job but who can safely do other work. When people are off work because of injuries, effective action is taken to assist in their rehabilitation and to ensure their early **return to work**. Line management and supervision take responsibility for these efforts, and they are conducted in a thorough but sympathetic manner.

   In this context, please rate the effectiveness of the modified duty and return-to-work initiatives of your organization.

   ☐₁ Excellent
   ☐₂ Good
   ☐₃ Satisfactory
   ☐₄ Poor
   ☐₅ Very poor

The results are given in Fig. 6-36. The question was added after the research project was partly completed, and data were obtained from

Modified Duty & Return-to-Work Systems			
	% Who Rated Modified Duty & Return-to-Work Systems As ...		
	Excellent	**Good**	**Satis-factory**	**Poor or Very Poor**
Best Result	70	28	2	0
Safe Co. Avg.	53	29	11	7
Unsafe Co. Avg.	13	20	23	44
Worst Result	13	20	23	44

Figure 6-36   In the safe companies, people have a positive view of the modified duty and return-to-work systems (Q17).

Modified Duty & Return-to-Work Systems				
	% Who Rated Modified Duty & Return-to-Work Systems As ...			
	Excellent	Good	Satis-factory	Poor or Very Poor
Best Result Man. & Supervisors	71	29	0	0
Best Result Workers	67	33	0	0
Worst Result Man. & Supervisors	42	42	16	0
Worst Result Workers	8	17	21	54

Figure 6-37   In the safe companies, workers and management agree that their return-to-work systems work well; in the one unsafe company polled, they disagreed (Q17).

only three companies with very good safety and from only one company with poor safety. Nonetheless, the results were generally as expected. Even in the very safe companies, this system did not get unanimous approval.

In Fig. 6-37, the perception of management is compared to that of workers for the company with the "best" and the company with the "worst" results.

There was a great difference between the perception of people in the company with the best results (S&C Electric) and the people in the company with the worst results. At S&C, there is a high level of trust between workers and management. There all workers viewed the return-to-work system as at least good, and two-thirds rated it as excellent.

Comments on Question 17

The results for this question illustrated two important insights that apply to the perceptions about the return-to-work system but are also typical of many of the questionnaire survey results. In companies with very good safety, the ratings were high and there was little difference between the views of workers and of leaders. In companies with poor safety, the ratings were much lower and there was a big difference between the perceptions of workers and those of management.

QUESTION 18: OFF-THE-JOB SAFETY

The organization has a responsibility to promote "off-the-job" safety as well as safety in the workplace, to help make safety "a way of life". (Belief 17)

If excellence in safety requires the vigilance that comes from an enduring mind-set of people, how can we expect to have excellence in safety at work and ignore safety in the rest of our lives?

Because of concern about intruding in the personal lives of employees, many companies are wary about taking a proactive approach to off-the-job safety. They should not be wary. Off-the-job injuries can be a major economic loss to the company as well as a personal tragedy and an economic hardship to the individual, and the safety capability of the work organization can be a real asset in improving off-the-job safety.

The safest companies understand that safety awareness must be seamless and that they have a responsibility to help their people (really to help each other) to become as safe off the job as they are on the job. They keep thorough records of off-the-job injuries so that trends can be observed. In a less direct sense than for workplace injuries, they take responsibility for reducing the off-the-job injury frequency of their people.

Off-the-job safety programs must obviously be less direct and slanted towards providing education and advice. However, the key to removing concern about interference in the private lives of employees is to have the off-the-job programs conducted mainly by working-level employees, with the full support of the company in time and resources. Many of the programs should also be open to family members (safe driving courses are an example).

The Results (Question 18)

18. To what extent is **"off-the-job" safety** dealt with in the workplace safety program of your organization? Check only one answer.

   $\square_1$ Off-the-job safety is an important part of our safety program. We keep statistics on off-the-job injuries, have an off-the-job safety committee, programs to promote safety in the home, safe driving off the job, etc.

□₂ Off-the-job safety is not a formal part of our workplace safety program but aspects of it are sometimes included in safety meetings, etc.

□₃ Off-the-job safety is not part of our workplace safety program.

The results in Fig. 6-38 show dramatic differences between the off-the-job safety efforts of the very safe companies and those with very poor safety.

Off-the-job safety is an important part of the safety efforts and culture of the very safe companies. It might be expected that people would be at greater risk of injury at work than off the job. There are few statistics available to determine whether this is true or not. It is known that among the very safe companies, employees are safer *at work*. At DuPont off-the-job safety has been a formal responsibility of managers and supervisors for decades, and it shows in the results. DuPont employees are much safer off the job than employees at most companies are at work. Their lost work injury frequency off the job for the 5 years from 1993 through 1997 was 0.35 (on the job it was 0.03). In comparison, the on-the-job LWIF for the five companies with poor safety was 20!

As expected, in the companies with very poor safety, respondents reported that there is little attention to off-the-job safety.

Off-the-Job Safety		
% Who Said Off-The-Job Safety Is ...		
Integral Part Of Program	**Present But Not Formal Part**	**Not Part Of Safety Program**
Best Result	99	0	1
Safe Co Avg.	52	43	5
Unsafe Co. Avg.	10	35	55
Worst Result	7	7	86

Figure 6-38   Off-the-job safety is stressed in the very safe companies, not in the companies with poor safety (Q18).

QUESTION 19: RECOGNITION FOR SAFETY PERFORMANCE

Recognition for safety achievement and celebration of safety milestones provide strong reinforcement for the organization's commitment to excellence. (Belief 18)

Just as in other areas of management, recognition can be a powerful way to reinforce the organization's values and goals. The standard is "a rich and varied array of ways to recognize individual and group achievements." Not rich in a monetary way but rich in the variety and scope of the recognition practices, a standard not often achieved.

There is disagreement on whether there should be "rewards" as well as recognition. Most of the very safe companies reinforce recognition with tangible items, for example, by giving members of a group a memento to mark their success. However, most reject the idea of monetary rewards for safety achievements, for "paying individuals for working safely."

Recognition for reaching a statistical milestone should, for the most part, be for group, department, or company achievements, not for individual safety. It seems inappropriate to give prizes to individuals or small groups for going injury-free for a year if others in their organization are injured or the safety goals are not achieved. Most employees in safe companies go their whole career without injury. The stress should be on teamwork to improve safety for all. Individuals should always be recognized for unusual contributions to the safety of others (e.g., development of a valuable safety practice).

Recognizing people in the presence of their peers or of their family members has an unusually powerful effect. This is often forgotten in safety as it is in other areas.

The Results (Question 19)

19. Indicate the extent to which achievements in safety are **recognized** and good safety performance is celebrated in your organization. Check the one answer that best represents your opinion.

   $\Box_1$ Thorough and extensive
   $\Box_2$ Frequent
   $\Box_3$ Some
   $\Box_4$ Little
   $\Box_5$ None

Workers' Views on Recognition for Safety Achievement				
	% Of *Workers* Who Said Recognition Is ...			
	Thorough Extensive	Frequent	Some	Little or None
Best Result	100	0	0	0
Safe Co. Avg.	51	37	9	3
Unsafe Co. Avg.	6	14	22	58
Worst Result	8	8	4	80

Figure 6-39 Recognition for safety is frequent in the very safe companies, virtually unknown in those with poor safety (Q19).

The results had a wide spread, even among the companies with very good safety. In the companies with poor safety, few people perceived recognition to be at a satisfactory level. The perception of workers is given in Fig. 6-39.[xii]

At the Fort Frances mill of Abitibi-Consolidated, recognition is heavily stressed, and it is reinforced with substantial monetary rewards for mill, department, and individual achievements in safety. The monetary reward underlines the company's serious intent to avoid injury. At S&C Electric, recognition gets equally high marks. It consists of a broad array of ways to recognize individuals, groups, and the achievements of the company as a whole. In both cases, the record of the organization in avoiding injury to anyone is the main criterion.

QUESTION 20: EMPLOYING THE BEST SAFETY TECHNOLOGY

The containment of hazards by integrating leading edge safety technology into the design and operation of facilities is essential for outstanding safety. (Belief 19)

In this book, it is stressed that good safety is built on management commitment, line ownership, and individual safety awareness developed largely through involvement. This does not mean that utilizing the best

---

[xii] This "worst result" is not the one given in Appendix F. Fig. 6-39 gives the worst *workers* opinion.

safety technology is not important. Rather, if these "soft" elements are in place, the continuous seeking out and application of the best technology, hard and soft, will be an automatic result.

Finding the best technology is not just the province of safety specialists. Line management, including workers, should take the lead in this endeavour. In one company with average safety that suffered from a lack of management commitment and involvement, an imaginative benchmarking initiative was carried out entirely by safety specialists. They missed a golden opportunity to gain commitment by involving the key managers in seeing firsthand how some very safe companies managed. This company's safety department wondered why their good benchmarking report received little attention. Safety management is still shot through with such elementary bad management practice, much as quality management used to be.

The Results (Question 20)

20. How do you rate the safety of the **physical facilities** in your workplace (machinery, equipment, etc.)? Check the one answer that describes your assessment.

☐₁ Excellent
☐₂ Good
☐₃ Satisfactory
☐₄ Poor
☐₅ Very poor

The results, given in Fig. 6-40, showed a clear separation between the perception of employees in very safe companies and in those with poor safety.

Almost all of the respondents in the safe companies considered the safety of their facilities to be good or excellent. One-half or more of those in the unsafe companies rated their facilities as only satisfactory, or poor or very poor. In the results from the unsafe companies, workers commented on unsafe situations in their workplaces:

· The area where we mix our paint for spraying is completely unventilated. This makes mixing paint quite unpleasant.
· When it rains, it rains in the warehouse. Many of the aisles are wet and slippery. An accident is waiting to happen with forklift drivers flying around corners.

The Safety of Facilities & Equipment				
	% Who Said the Safety of Facilities & Equipment Are ...			
	Excellent	Good	Satis-factory	Poor or Very Poor
Best Result	65	32	3	0
Safe Co. Avg.	44	48	7	1
Unsafe Co. Avg.	5	46	31	18
Worst Result	0	20	40	40

Figure 6-40   In the safe companies, people judged the safety of facilities to be at least good. Not in those with poor safety (Q20).

- I was personally forced to run my machine without guards on two cutters for over 4 months. I had to make one with wood and clamp it on with c-clamps.

Comments on Question 20

Injuries occur from the interaction of people with their physical environment. Thus it might be asked why the questionnaire does not deal at more length with the state of equipment, safety devices, and physical guards against injury. The rationale is that the most important factors are the commitment of management, the way that line management takes responsibility for safety, and the safety awareness of workers. If these are firmly in place, all the best safety technology in equipment, systems, and management processes will be utilized and continuously improved.

QUESTION 21: MEASURING AND BENCHMARKING SAFETY PERFORMANCE

Comprehensive, up-to-date safety statistics, communicated to all, are a cornerstone of safety management. Benchmarking against the best will help improve safety. (Belief 20)

"We manage what we measure" is a useful credo. The very safe companies record a wide array of safety statistics—lost work injuries, restricted work injuries, medical treatment cases, first aid cases, incidents with potential for injury, and so on. Their safety goals and annual objectives are related to these measurements. They regularly bench-mark their performance against that of the safest companies in their industry and beyond. The safety information is widely and frequently communicated to the workforce.

Injury statistics are lagging indicators of what has happened. The very safe companies also endeavour to find coincident and leading indi-cators of the state of safety. These include incident frequencies, quan-tified workplace audits, and systems to assess employee attitudes to safety (e.g., perception surveys). In contrast, companies with poor safety records often keep barely adequate lagging statistics, and often their employees are poorly informed of how their company's safety compares to that of others.

Question 21 does not deal directly with assessment of the quality of the safety statistics and benchmarking; rather, it assesses the knowledge of employees about them.

The Results (Question 21)

21. Check the one answer that best represents the extent of your **personal knowledge of the safety performance** of your organization:

$\square_1$ I know our safety goals and our up-to-date performance. I know how our performance compares with that of other companies in our industry.

$\square_2$ I know our safety goals and present performance but do not know how our performance compares with that of other companies.

$\square_3$ I am only generally aware of our safety goals and how we are doing in safety. I do not know how we compare with others.

$\square_4$ I do not know our safety goals. I am not familiar with how we are performing in safety. I do not know how we compare with others.

Figure 6-41 shows the state of knowledge of *workers* about their organization's goals, performance, and comparison to peer companies.

In the very safe companies, the level of knowledge was high. In the companies with poor safety, the level of knowledge was low. The Fort Frances mill of Abitibi-Consolidated publishes a daily news sheet. Safety is the first item, with the up-to-date safety record superimposed on the safety triangle at the top of the sheet. The news sheet includes

Workers' Knowledge of Safety Performance				
	% Of *Workers* Who Said Their Knowledge of Safety Performance Was ...			
	Full	Fair	Only General	Little or None
Best Result	82	18	0	0
Safe Co. Avg.	58	35	6	1
Unsafe Co. Avg.	6	35	47	12
Worst Result	8	17	46	29

Figure 6-41   Workers in safe companies know much more of their company's safety performance than in the unsafe companies (Q21).

lots of things of interest to people in the mill—plant sports team results, plant activities, news about plant people—so it, and therefore the safety information, gets widely read.

Comments on Question 21

Rather than asking directly about the quality of the statistics and benchmarking, this question probed the knowledge of respondents about the organization's goals and performance. The rationale was that the best overall test of the quality of the data collection and benchmarking would be how well these things were known by employees. The answers to the question matched the safety performance of the responding companies.

QUESTION 22: THE SAFETY ORGANIZATION

The safety organization is a valuable asset in attaining excellence in safety. It should be led by the line organization with broad participation by the entire workforce, particularly those at the working level. (Belief 24)

The safety organization includes the senior management safety committee, the committees that monitor and improve systems on behalf of the workplace (for example, a rules and procedures committee), and task forces or teams set up for specific purposes. It does not include the safety department or safety specialists, which are covered by question 23.

The safety organization should be operated by line management with broad participation of the entire workforce. The senior safety committee or team should be chaired by the senior leader in the workplace, for example, the plant manager or the president, and should be made up of all the managers who report to the leader and the main committee chairpersons. In the very safe companies, attendance at the regular meetings of this central safety committee has high priority and a very compelling reason is needed for a member to be absent. This reinforces the number one priority given to safety.

The Results (Question 22)

22. How do you rate the effectiveness of the **safety organization** in your workplace (the managers' safety committee, the JOHSC, other safety committees, the safety systems, structures and procedures, etc.)? Check only one answer.
    $\square_1$ Excellent
    $\square_2$ Good
    $\square_3$ Satisfactory
    $\square_4$ Poor
    $\square_5$ Very poor

The perception of the quality of the safety organization is given in Fig. 6-42.

The respondents in the very safe companies rated their safety organization as good or excellent. In the companies with poor safety, the rating was at best satisfactory.

Rating of the Safety Organization			
	% Who Rated the Safety Organization ...		
	Excellent	Good	Satis-factory	Poor or Very Poor
Best Result	57	38	5	0
Safe Co. Avg.	35	53	11	1
Unsafe Co. Avg.	5	35	41	19
Worst Result	0	27	43	30

Figure 6-42 The safety organization was rated much higher in the safe companies than in those with poor safety (Q22).

QUESTION 23: THE SAFETY DEPARTMENT—SAFETY SPECIALISTS

Safety specialists can provide valuable assistance to the safety organiza-tion. They must avoid taking responsibility for managing safety or accept-ing accountability for results; these lie with the line organization. Rather than doing the work themselves, they should facilitate involvement of the workforce. (Belief 25)

Safety specialists can help an organization to achieve better safety. However, their presence can sometimes be counterproductive, provid-ing an excuse for management to improperly relinquish its responsi-bility and for worker involvement to be minimal.

Rather than doing the work themselves, safety specialists should con-centrate on facilitating the involvement of management and workers in doing the safety work. This is a hard role to fill and requires a mind-set and a skill not usually found in people in the safety specialist position.

Safety specialists can play a valuable role in ensuring that the safety systems (such as statistical record systems) are in good condition and kept up to date. They can provide expert knowledge on government safety regulations. In larger organizations, they can facilitate the trans-fer of technology from one department to another. In general, though, they must stay well in the background and avoid doing the work or taking the responsibility that properly belongs to the line.

The Results (Question 23)

23. How do you rate the effectiveness of the **safety department** in your organization (the safety supervisor, the safety advisors, etc.)? Check only one answer.

☐$_1$ Excellent
☐$_2$ Good
☐$_3$ Satisfactory
☐$_4$ Poor
☐$_5$ Very poor

The respondents in the very safe companies generally gave their safety departments good marks. In the companies with poor safety, respondents rated their safety departments as poor (Fig. 6-43).

Milliken has no safety specialists in its workforce. Line managers,

Rating of the Safety Department			
	% Who Rated the Safety Dept. ...		
	Excellent	Good	Satis-factory	Poor or Very
Best Result	45	52	3	0
Safe Co. Avg.	33	49	16	2
Unsafe Co. Avg.	9	33	36	22
Worst Result	0	8	40	52

Figure 6-43   The safety department was rated much higher in the safe companies than in those with poor safety (Q23).

supervisors, and workers do the safety work. The goal is for everyone to be directly involved in specific safety activities—committees, audit-inspections, etc. Milliken is a long-established company with management depth and continuity, and this practice would not necessarily work for other companies. However, Milliken believes that this is an important reason for their excellent safety record (averaging less than 0.1 LWIF).

We should not attribute too much to criticism of the safety specialists in the responses to this question. It is easy to focus frustration about safety on the safety organization or the safety department. The essence of the internal responsibility system is the ownership of the safety agenda by line management, from the CEO to the individual.

QUESTION 24: SATISFACTION WITH THE SAFETY PERFORMANCE OF THE ORGANIZATION

The last question probed the respondents' satisfaction with the safety performance of the organization. This question did not relate to any specific belief.

24. To what extent are you personally **satisfied with the safety performance** of your organization? Check only one answer.
    $\Box_1$ Very satisfied
    $\Box_2$ Moderately satisfied
    $\Box_3$ Neither satisfied nor dissatisfied
    $\Box_4$ Moderately dissatisfied
    $\Box_5$ Very dissatisfied

Satisfaction with Safety Performance				
	% Who Said That They Were ...			
	Very Satisfied	Moderately Satisfied	Neutral	Dis-satisfied
Best Result	78	22	0	0
Safe Co. Avg.	53	43	2	2
Unsafe Co. Avg.	10	40	21	29
Worst Result	10	10	27	53

Figure 6-44 Respondents in the unsafe companies reported a surprising level of satisfaction with the safety performance (Q24).

Satisfaction is a subjective thing, as can be seen in the results in Fig. 6-44.

Despite the virtual elimination of major injuries (less than 0.1 LWIF over 5 years), respondents at the very safe companies were not totally satisfied. They work to a very high standard. The plant manager and the senior managers at one DuPont plant that had not had a lost work injury for many years reported that they were only moderately satisfied. At the time of the survey in mid-year they had experienced just one medical treatment case, but their goal for the year was zero.

Despite living with a terribly high injury frequency, people at the companies with poor safety were not as dissatisfied as one might expect. How can more than 50% of the respondents in these compa-

Satisfaction with Safety Performance Unsafe Companies				
	% Who Said That They Were ...			
	Very Satisfied	Moderately Satisfied	Neutral	Dis-satisfied
Managers	10	50	14	26
Supervisors	4	66	15	15
Workers	16	30	25	29
All	10	40	21	29

Figure 6-45 Workers in the unsafe companies were somewhat less satisfied with safety than were their leaders (Q24).

nies be satisfied with their safety? Their standards are obviously low. This was a bad result for them because dissatisfaction is a spur to continued efforts to improve.

Among the companies with poor safety, workers were somewhat less satisfied with safety performance than managers were, but it is difficult to understand why they indicated any level of satisfaction (Fig. 6-45).

BELIEFS AND PRACTICES FOR WHICH NO QUESTIONS WERE DEVELOPED

Although the questionnaire covers almost all of the important issues in safety management, there are some beliefs and practices that cannot be assessed well with this technique. The concepts behind these beliefs were discussed in the interviews with company leaders of the very safe companies (Appendix C).

Belief 5: Good Safety is "Mainly in the Head"

Good safety is "mainly in the head." Most injuries and incidents occur because of inattention, not because of lack of knowledge or for physical reasons. People take risks because they believe that they will not be hurt.

Discussion of this belief leads to better understanding of and insights about safety management. However, it does not lend itself to a direct question. This concept pervades all of the safety culture of the very safe companies and results in concentration on attitude and behaviour, not just on systems and physical things. The questions about the priority given to safety, about business-safety interactions, and about involvement relate to this belief. The practices also build on the philosophy of influencing attitude through training and involvement in safety activities.

There were no questions designed to refer to the following three practices. These practices would not be uniformly well known by the cross section of people involved in the surveys. However, the managers in the five very safe companies were questioned about the practices in the interviews with them.

Practice 21: Hiring for Safety Attitude

Safety can be enhanced by hiring people with good safety values and attitudes.

This practice is based on the concept that safety is behavioural and that bad habits, once learned, may be hard to change. Thus knowledge of and attitudes towards safety should be one of the subjects in interviews and integrated into the testing of prospective employees. Some companies have adopted the practice of engaging candidates with existing employees in team problem solving as part of the hiring process, enabling assessment of attitudes and interpersonal skills as well as ability. This affords an opportunity to investigate the candidate's knowledge of and attitude towards safety. This technique is used by at least two of the very safe companies surveyed here.

Practice 22: Safety of Contractors and Subsidiaries

Contractors and subsidiaries, including foreign operations, must work to the same safety standards as the company.

The very safe companies accept the principle that they have a responsibility for the safety of contractors. In these companies, there are formal programs to see that only very safe contractors are hired and that employees of contractors work to the same standards as their own employees while in the company's facilities.

The standards for safety in DuPont, Milliken, and Shell are the same worldwide for all foreign and subsidiary operations. (S&C Electric is a smaller company with its main operations in the US and Canada. Abitibi-Consolidated is a recent assembly of pulp and paper companies and mills.)

Practice 23: Involvement in Community and Customer Safety

Excellence in safety will lead naturally to involvement in and leadership in community and customer safety. This will be valuable to the organization and to the community and customers.

In the very safe companies, employees at all levels are deeply involved in local community safety and fire protection activities and in national and international safety organizations.

7

THE SAFETY MANAGEMENT APPROACHES OF FIVE VERY SAFE COMPANIES

The central purpose of the research project was to assess in depth how a few companies achieve safety results far better than the average. The safety questionnaire was an important part of the research. It provided a unique way to measure the key parameters of safety management in the five very safe companies and in the five companies with very poor safety. In Chapter 6, the analysis of the results of the questionnaire survey are presented. The research also included extensive on-site research in the five very safe companies—head office and site visits, interviews with executives and managers, focus groups of workers and supervisors, and gathering of extensive information on the companies and their safety management methods. These comprehensive investigations augmented and deepened the understanding that emerged from the analysis of the questionnaire data.

The intent of the write-ups that follow is to meld all of the research findings into a coherent story on each of the five companies, a story that captures the unique management approaches that they have used to attain enduring excellence in safety.

7.1 ABITIBI-CONSOLIDATED, FORT FRANCES MILL: SAFETY EXCELLENCE IN PULP AND PAPER PRODUCTION

7.1.1 Introduction

Forest products is the largest industry in Canada and is a major exporter and a large employer. The pulp and paper sector is a large part of this industry. It is characterized by complex and hazardous operations. A company with excellent safety from this sector was sought for the research project.

For many years the magazine *Pulp & Paper Canada*[i] has published a list of the "Safest Mills in Canada." Providing the data is voluntary, and some companies or mills do not participate. However, the list has included most of the safest mills, classified as small, medium, or large mills by the hours worked per year.

In this database, no *company* has stood out in safety. It is difficult to follow companies because of the many mill shutdowns, sales of companies and mills, and mergers. Through the 1990s, the safest large *mills* were the two at Fort Frances and Kenora, Ontario that were formerly part of Boise-Cascade. The history of these mills illustrates the problem of corporate identification. After many years with Boise, in 1994 the two mills became the main assets of a Canadian company called Rainy River Forest Products. In 1996, Rainy River was purchased by Stone-Consolidated and became part of an eight-mill Montreal-based company. Then in 1997, Stone-Consolidated merged with Abitibi to become Abitibi-Consolidated, the largest producer of newsprint in the world.

Boise Cascade had always stressed safety, and the commitment was continued under Rainy River. For the 5 years 1993 from through 1997, the average LWIF of Fort Frances and Kenora combined was under 0.3, six times better than the average for the 20 large mills listed by *Pulp & Paper Canada*. However, this is not up to the standard for "world class," and the Kenora mill had two fatalities in the period.[ii] The record of the Fort Frances mill was much better. On the basis of these considerations, it was decided to include only the Fort Frances mill in the research.

The Fort Frances mill is located on the Rainy River in northwestern

---

[i] Published monthly by Southam Magazine & Information Group.
[ii] The author consulted for the Kenora mill in 1994–96. In this period, the mill effected a turnaround, reducing its lost work injuries from an average of 6 per year in 1991–94 to 0 in 1995–97 (25). The questionnaire used in this research, initially developed for consulting, was used in the Kenora work.

Ontario, just across from the mill of Boise Cascade in International Falls, MN. The two mills were built in the early 1900s to exploit the softwood on both sides of the border. Production began at Fort Frances in 1914. The mill is the largest employer in the region and has a reputation as a good place to work—safe, responsible, with well-paid jobs. Three unions represent the mill workers. The mill has had a good record for labour harmony but in 1998 was part of a protracted company-industry strike. Participation of workers in safety, in quality, and in other areas has steadily increased.

In the late 1990s, the Fort Frances mill capacity was about 400,000 tonnes per year. It employed about 800 people, not including woodland workers, who are employees of contract pulpwood suppliers. The mill specializes in high-quality, uncoated groundwood printing paper for magazines, advertising materials, directories, and books. Market pulp is also produced for manufacturers of newsprint, tissue, and office paper.

The mill has a reputation for quality and is considered one of the top specialty mills in North America. In a survey of customers for uncoated groundwood paper, Abitibi-Consolidated was rated as the market leader. Many of the comments that related to speciality papers praised the products of the Fort Frances mill.

According to Jim Gartshore, manager of the Fort Frances Division, "the mill has been profitable." No doubt this is related to its emphasis on technology and quality. Its superior record in safety has thus been part of an overall standard of business excellence.

7.1.2   The Safety Record of the Fort Frances Mill

The statistical performance of the Fort Frances mill, derived from its records and from those published in *Pulp & Paper Canada*, is summarized in the table in Fig. 7-1.

Among the large mills, Fort Frances has been in a class of its own. In the 1990s, no other mill approached its level and consistency of safety performance. The closest was its sister mill at Kenora, which had an average LWIF for 1993–1997 of 0.4. (The Kenora mill has improved greatly in the last few years. It had no lost work injuries in 1995, 1996, 1997, or 1998.) None of the medium or small mills was close to Fort Frances in performance.

There is no complete record of the total injury frequency for Canadian mills. The Ontario safety association for the sector reports total injuries for *Ontario* mills. The Fort Frances and Kenora mills have consistently led in the ranking.

Although the safety performance of the Canadian pulp and paper

The Safety Record of the Fort Frances P&P Mill										
	'90	'91	'92	'93	'94	'95	'96	'97	'98	Avg. 1993-7
Fort Frances										
LW Injuries (No.)	5	3	1	3	1	0	0	0	0	1
LWIF	0.65	0.38	0.13	0.40	0.13	0	0	0	0	0.11
TRIF	3.40	3.90	2.50	4.90	2.1	0.9	1.6	1.6	~1.6	2.0
Ranking	2	1	1	1	2	1	1	2	~	1
Large Mills										
No. of Mills	29	27	20	22	20	20	21	19	15	
LWIF, Avg.	4.1	4.2	2.7	2.2	2.0	1.6	1.4	1.1	~1.0	1.7
TRIF, Avg.	16	16	13	12	11	10	9.6	8.1	~	10.1
The ranking is the placing of Fort Frances among the large mills in LWIF. The TRIF for the large mills is for *Ontario* mills as reported by the Ontario safety association for the sector (TRIF not reported by Pulp & Paper Canada). In 1998, Abitibi mills were not included in the data since they were on strike for 6 months.										

Figure 7-1 A fine safety record but total injury frequency not world class.

companies has improved substantially, this industry has not been in the top rank compared with other industries. The dozen large chemical companies averaged 0.5 in LWIF and 2.5 in TRIF for 1993–1997.[iii] The safety performance of the pulp and paper mills was probably worse than given here, because listing in *Pulp & Paper Canada* is voluntary and some poor performers do not participate. In contrast, filing safety data is a condition of membership with the CCPA.

Thus, in the context of world class safety, the Fort Frances mill, although very good, had not been outstanding. Some manufacturing *companies* have achieved average LWIFs below 0.1. Within these companies, these are *plants* that have had no lost work injuries for many years. These companies also have had lower TRIFs, a better indication of safety performance, in some cases well below 1.0. Despite these qualifications, Fort Frances has been strikingly better than its peers.

7.1.3   The Safety Survey at the Fort Frances Mill

The safety survey at the Fort Frances mill included these elements:

· Examination of mill safety data and other documents
· Safety questionnaires completed by employees

[iii] Chemical company data from Canadian Chemical Producers' Association (CCPA).

Responses to Questionnaires	
Group	No. of Questionnaires
Corporate Management	--
Other Managers	*10*
Total Management	10
Supervision	10
Professional	5
Working Level	11
Undesignated	4
TOTAL	40

Figure 7-2 Responses to questionnaires, Fort Frances.

- Interviews with the mill manager and the managers of human resources and of health and safety
- Visits to the mill and discussions with managers and workers
- Discussion of the questionnaire results with two groups of employees

The results of the questionnaire were discussed with two groups at the mill, one a group of managers and one involving a cross section of people. (This was before the focus group system, used later, was formalized.)

Questionnaires were distributed to 65 people, including the mill management group, other managers and supervisors, professional staff, and workers (Fig. 7-2). Of the 42 returned, 40 were usable. It was disconcerting that only about half of the workers returned completed questionnaires. It is likely that the four questionnaires turned in without the job categories being designated were from the working-level group.

Some of the results that appear key to the success of Fort Frances and some that indicate where further improvement might be sought are given below. The questionnaire results are augmented with insights from the focus groups and the interviews.

7.1.4 Insights from the Fort Frances Safety Survey

7.1.4.1 *Commitment of Management to Excellence in Safety.* To achieve its level of safety performance, the Fort Frances mill obviously does most things right. It was not always thus, however. Although Boise Cascade had always stressed safety, in the 1980s the Fort Frances mill's

performance was good but not outstanding. In 1990, Dave Blenke from Boise Cascade (US) was appointed mill manager. He stayed only a short time but had an important impact on the safety culture. The management team picked up and extended his enthusiasm for excellence in safety. Len Robinson, the manager of human resources, was in the mill through this period:

> Prior to 1990, in many cases production came first. Not miles ahead, but the feeling was that at the end of the day you were judged by how much saleable product was turned out. If safety was good, fine. Then along came this new guy (the new mill manager) with a real passion for it. He was really revved up. You could tell that he was serious about this stuff. And he infected us. We were good but he got us much better. Put the focus on safety. The passion was there. It hasn't affected business performance. This was and still is a very successful mill.

The mill manager chairs the management safety committee. The mill manager and line managers said that they spend 10–15% of their time on safety. Safety performance is a part of their incentive compensation. The mill manager holds the managers and supervisors responsible for chairing investigations, for training, although there is a training section to help, and for meeting specific objectives in safety.

The safety drive of the early 1980s brought focus to an already good safety culture. In the late 1990s, there was a strong commitment to continuous improvement, towards eliminating all injuries. As one manager put it, "We were the safest mill in Canada with a total injury rate of 4. Now we are still the safest with a rate approaching 1."

The Fort Frances respondents reported that they *individually* give high priority to safety (question 1). However, they were not so sure that others did (question 2) (Fig. 7-3). Most respondents said that workers give high priority to safety. However, workers did not perceive that supervisors or managers rank safety first. This is not a world-class result.

Workers thought that managers give equal priority to costs and safety. They thought that supervisors give higher priority to production volume than to the other three factors. The divergence in perception between management and workers is a negative factor.

7.1.4.2 *Safety Values.* Most of the important safety values were present in the culture of the mill. However, there was uncertainty as to whether there was a *written* set of safety values and whether they were

The Priority People Think Others Give to Safety					
	% Who Think Others Rank Safety First				
	Abitibi-FF	Best Result	Safe Co. Avg.	Unsafe Co. Avg.	Worst Result
Managers' & Supervisors' View of Workers' Priority	75	96	81	39	41
Workers' View of Managers' & Supervisors' Priority	32	87	66	19	8

Figure 7-3 Workers at Fort Frances did not think that management gives high priority to safety (Q2).

The Influence of Safety Values			
	% Who Said Safety Values Have ...		
	Important Influence	Some Influence	Little or No Impact
Abitibi-FF	45	27	28
Best Result	88	10	2
Safe Co. Avg.	67	22	11
Unsafe Co. Avg.	12	28	60
Worst Result	3	26	71

Figure 7-4 Safety values were not perceived to be as influential at Fort Frances as in other safe companies (Q7).

a strong influence (Fig. 7-4). This was the lowest answer of the five safe companies.

Shortly after this survey was completed, the mill management team met to discuss the mill's safety values. Then they began a process to communicate them and entrench them in the mill's safety culture. The speed with which the management moved to fill this perceived gap is an indication of their commitment to excellence in safety.

Unlike many workplaces, at Fort Frances there is a strong belief that excellence in safety contributes to excellence in quality and in other business parameters. On question 4 on this topic, the response was close to the best result (Fig. 7-5).

The Effect of a Drive for Safety on Business			
	% Who Said a Drive for Safety Affects Business Performance in Following Way		
	Very Helpful	Positive	Neutral	Makes Difficult, Weakens
Abitibi-FF	74	20	3	3
Best Result	73	26	0	1
Safe Co. Avg.	67	29	1	3
Unsafe Co. Avg.	32	40	15	13
Worst Result	24	42	12	22

Figure 7-5 Fort Frances respondents thought good safety has a very positive influence on business results (Q4).

Similarly, most respondents thought that there was no limit to the level of safety to which they could aspire and still not reach a cost-benefit breakpoint (question 5). In these results, the mill was among the best of the companies surveyed.

7.1.4.3 The Extent Management Held Accountable for Safety. The Fort Francis answers were right at the average of the results for the five very safe companies in this important parameter.

7.1.4.4 Involvement in Safety Activities. The managers estimated that 10–15% of the workers were involved in safety activities beyond their job. Although involvement had increased, it is higher in the safest companies. Involvement in audits was not high. Some workers saw auditing as a threat, particularly auditing that deals with safety awareness. The mandated Joint Health and Safety Committee was dysfunctional at the time of the survey. Two of the four union locals were not participating.

7.1.4.5 Safety Training. The mill respondents gave a good rating to the extent of training—in the middle of the responses from the five very safe companies. One novel initiative is a one-day refresher course in safety conducted by workers with the help of the training department. It includes everyone in the mill in diagonal slice groups every year. It is a good opportunity to stress areas that need focus. Before visitors enter the mill, they must take a short but thorough safety course. They are given a certificate good for 6 months. This excellent system is indicative of the mill's comprehensive approach to safety.

7.1.4.6 Safety Rules and Their Enforcement. The mill respondents rated the quality of their safety rules as fairly good and felt that they were generally followed. In each of these questions, the Abitibi results were lower than those of the other very safe companies but not significantly so.

In question 14, the respondents were asked to indicate the extent to which safety rules are enforced. The results are shown in Fig. 7-6.

There was a wide range of responses to this question, even in the very safe companies. Many managers and supervisors are reluctant to accept the ideal standard that all infractions result in some action, if only a conversation. The Fort Frances results were the lowest of the very safe companies by a considerable margin. The mill had a general policy for disciplinary action. However, disciplinary action for safety infractions was not well specified, and the results show that it was not being applied consistently.

7.1.4.7 Other Safety Practices. The mill keeps excellent records of injuries and incidents. This is an important initiative—"What gets measured gets done." The mill stood out in the knowledge that its people had of the mill's safety goals, of its performance, and of how it compared with the safety of other pulp and paper mills (Fig. 7-7).

The answers from the mill respondents were the highest of the com-

Enforcement of Safety Rules				
	% Who Said Disciplinary Action Taken for ...			
	All Infractions	**Serious Infractions**	**Not Consistent Arbitrary**	**Seldom Taken**
Abitibi-FF	15	25	13	47
Best Result	67	20	12	1
Safe Co. Avg.	45	27	11	17
Unsafe Co. Avg.	18	21	27	34
Worst Result	4	16	11	69

Figure 7-6 The responses indicated that the rules were not as well enforced at Fort Frances as at other safe companies (Q14).

Knowledge of Safety Performance				
	% With Knowledge of Safety Performance ...			
	Full Know-ledge	Fair Know-ledge	Only General	Little or None
Abitibi-FF	92	8	0	0
Best Result	92	8	0	0
Safe Co. Avg.	75	22	3	0
Unsafe Co. Avg.	13	40	38	9
Worst Result	10	23	44	23

Figure 7-7   Everyone at Fort Frances had very good knowledge of the mill's safety performance (Q21).

panies surveyed. Over 90% of the respondents said that they knew their mill's safety goals and performance and how they compared to other mills.[iv] One reason for this is their newspaper "Screenings," published daily by mid-morning. The first item is the safety statistics, overwritten on the safety triangle (Fig. 7-8). The paper is full of newsy items, so it gets read.

Another reason for the level of knowledge is that awards under their safety incentive program relate directly to safety performance. Fort Frances people had an excellent level of knowledge of the safety of their mill and of their industry. However, many were surprised to find that there were *companies* outside their industry with records superior to their *mill* record.

The respondents to the questionnaire had a strong belief in the importance of recognition for safety performance. The mill has a safety appreciation night with all the managers in attendance, and there are many other recognition initiatives. In this important parameter, the Fort Frances people registered the highest results of any of the companies surveyed (Fig. 7-9).

One unique feature of the safety program is the "Safety Points" award system. Employees accumulate points for mill-wide, departmental, and individual achievements. The points are redeemed for goods at local businesses. The rewards can be large—recent awards

[iv] Note that this refers to information about the local *mill*. The more difficult question posed to respondents at other companies was about *company* goals and performance.

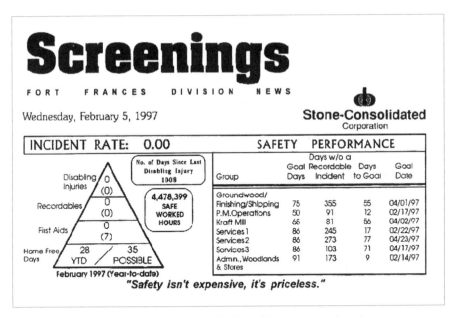

Figure 7-8   The first item in the daily newspaper is safety.

Recognition for Safety Achievement

	% Who Said Recognition Is ...			
	Thorough Extensive	Frequent	Some	Little or None
Abitibi-FF	92	8	0	0
Best Result	92	8	0	0
Safe Co. Avg.	55	35	9	1
Unsafe Co. Avg.	7	15	27	51
Worst Result	7	7	14	72

Figure 7-9   Recognition was judged to be at a high level (Q19).

have averaged $500 to $1000 per year for every person in the mill! Some employees believe that this approach is an important reason for their improvement in safety. The award program, started in the 1980s, was strengthened in 1993—and their safety has improved since then. Managers used to say that you couldn't buy safety, that the reward is to go home whole at the end of the day. Now they believe that the points system provides reinforcement to their safety effort. Knowledge of performance and recognition go together.

Although there is attention to off-the-job safety, it is not given as important a focus as would be expected, given the high rate of off-the-job injuries among Fort Frances people. Recently, however, the mill has intensified its attention to off-the-job safety.

7.1.5 The Keys to the Safety Success at Fort Frances

In an industry not noted for leading in safety, the Fort Frances mill stands out. The foundation of their success is the commitment of management to excellence in safety, evident in the words and actions of the present mill leaders. Jim Gartshore, the division general manager and mill manager said, "Our vision is to be the safest mill in the industry, to have zero lost time accidents, an incident rate of less than 1 and to continuously improve. Our vision is very important to our safety effort. We believe that all injuries are preventable. We believe that everything has to be investigated—we work our way right down to first aids and near-misses, to the bottom of the (safety) triangle."

7.2 DUPONT CANADA: ONE OF THE WORLD'S SAFEST COMPANIES

7.2.1 Introduction

DuPont Canada, one of the safest industrial companies in Canada, if not the safest, was the starting point for this research. It holds the Canadian plant and company records for the longest periods without a lost work injury. Its parent company, E I. du Pont de Nemours, is world-renowned for excellence in occupational health and safety.

The author is not unbiased about DuPont Canada, having spent much of his working life there. However, the results speak for themselves, and they speak of excellence in every facet of safety.

DuPont is one of the oldest industrial companies in the world, tracing its roots to black powder mills in Delaware in the early 1800s. It continues after almost 200 years to be among the largest and most successful companies worldwide, which speaks to its enduring strength and character. For many decades, wherever it operates around the world, it has been known for its uncompromisingly high standards in health and safety. Observers have said that DuPont has an obsession with safety, and indeed it is true. Legend has it that in the old days, when everyone lived within walking distance of work, the mill manager had to live between the black powder mill and the homes of the workers. The necessity for extraordinary precautions in gunpowder production became the basis for outstanding safety as the company expanded worldwide into chemicals, fibres, high performance materials and biotechnology.

In DuPont, the high value given safety and occupational health is not questioned. It is fundamental. If you can't deliver virtually injury-free performance in your organization, you won't be a manager for long. And although it is universally held in the company that excellence in safety is good for business, it is really driven more by human values. There is a strong belief that the two go together, that excellence in business is not possible without respect for the well-being of people. In safety, the DuPont Company sets the standard. There may be some as good, but none better.

DuPont Canada is itself a company of long standing, going back to before Confederation (28). For many years it operated as CIL, owned by DuPont and ICI with minority Canadian ownership. Since the 1950s, DuPont Canada has been 75% owned by DuPont, with much of the remaining 25% owned by Canadians. The company is technologically oriented and has contributed to its growth through research. Since the Free Trade Agreement, it has rationalized with its parent, producing specialized lines for world markets. Its sales volume is about $2 billion, and it employs about 3500 people.

7.2.2 The Safety Record of DuPont Canada

DuPont Canada inherited the high safety standards of its parent and has always had an excellent safety record. However, through the 1970s and 1980s, it did not measure up to the exceptional performance of its parent: it was usually in the lower half of the list compared to other units of the worldwide company.

In the mid-1980s, DuPont Canada undertook a major effort to transform itself from a solid but underperforming company to a pacesetter in business excellence. It shed weaker businesses, strove for world-competitive costs and quality, and reoriented its businesses towards global markets. It reduced the levels of management and intensified development of self-directed work teams. With the help of a strong economy, the company moved from strength to strength, improving its businesses so that most were fully world-competitive. In 1988, DuPont Canada won the Gold Medal for Productivity in the Canada Awards for Business Excellence. The company became highly profitable.

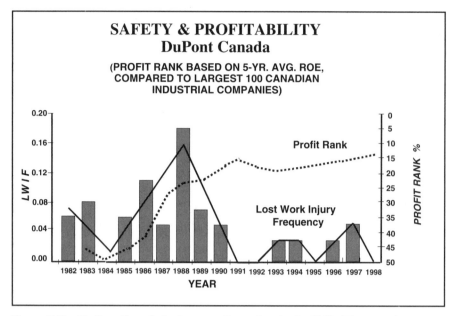

Figure 7-10   DuPont Canada had outstanding safety in the 1980s. The record was even better in the 1990s, with virtually no lost injuries.

During this transformation, an unexpected and serious decline in safety performance was encountered. The story of the resolution of this problem was told in *Business Quarterly* (3). Using a principle-centred management approach, the company recovered its excellent safety without sacrificing its progress in self-management or in other business-related factors. It has gone on to record better safety results than ever before, becoming one of the best performing units of DuPont world-wide. An update of the graph from the *Business Quarterly* paper is given in Fig. 7-10.

The 10-year safety record of DuPont Canada is given in Fig. 7-11. It is the only company of those studied that has readily available off-the-job safety statistics.

With this record, DuPont Canada can be said to have come close to eliminating lost work injuries, with only 5 in the 1993–1997 period in a population of 3500 people. DuPont would have had over 100 injuries had the injury frequency been at the average of the CCPA companies or over 500 if it had been at the average of Ontario manufacturing. In 1998, 1999, and 2000, DuPont Canada had three very good years, with no lost work injuries and near record lows in total recordable and off-the-job injuries. As of April 30, 2001 (and ongoing) the company had set new Canadian records for the time without a lost work injury—17

The Ten-Year Safety Record of DuPont Canada										
	'89	'90	'91	'92	'93	'94	'95	'96	'97	'98	Avg. 1993-97
DuPont Canada										
LW Injuries (No.)	3	2	0	0	1	1	0	1	2	0	1
LWIF	0.07	0.05	0	0	0.03	0.03	0	.03	.06	0	0.03
TRIF	0.61	0.86	0.52	0.46	0.37	0.56	0.36	0.26	0.36	0.33	0.38
OTJF (lost work)	0.54	0.77	0.60	0.64	0.41	0.18	0.27	0.28	0.29	0.17	0.29
CCPA										
LWIF	1.1	1.2	1.1	0.79	0.74	0.69	0.58	0.58	0.65	0.53	0.65
TRIF	3.2	3.2	3.7	3.1	2.9	3.0	2.4	2.8	2.7	2.1	2.8
The CCPA data is the average for the 60-odd companies of the Canadian Chemical Producers' Association.										

Figure 7-11 The excellent all-around safety performance record of DuPont Canada.

million hours for the Maitland plant and 26 million hours for the company.

7.2.3 The Safety Survey at DuPont Canada[v]

The safety survey at DuPont Canada included these elements:

1. Examination of company safety data and other documents
2. Questionnaires completed by 110 people, 43 people at the Whitby, ON. plant, 54 at the Maitland, ON. plant, and the 13 members of the senior corporate executive group
3. Research at the Whitby and Maitland plants:
   - Questionnaires completed by a cross section of employees, including all of the plant management teams
   - Tours of plants and discussions with workers and others
   - Interviews with plant managers, HR managers, and safety supervisors
   - Discussions with groups of managers and supervisors (this work was done before the use of focus groups was formalized)
4. Research at the Corporate Executive level:
   - Questionnaires completed by the CEO and his team (13)
   - Interviews with the CEO and the Vice President of Human Resources

[v] In the safety survey data, DuPont means DuPont Canada unless otherwise noted.

Responses to Questionnaires	
Group	**No. of Questionnaires**
Corporate Management	13
Other Managers	16
Total Management	29
Supervision	17
Professional	11
Working Level	53
Undesignated	--
TOTAL	110

Figure 7-12   Responses to questionnaires at DuPont.

The DuPont questionnaires were completed early in the research project. When new questions were added, a resurvey was undertaken in early 1998, just at Whitby and just for the three new questions (question 9, part 2 on involvement, question 12 on safety meetings, and question 17 on return-to-work systems). These data are included with the other DuPont data (Fig. 7-12).

Whitby is a non-union plant with about 300 people, producing specialty packaging films. It has an exceptional safety record, having gone more than 10 years without a lost work injury and having no recordable injuries in 1997 and 1998. The Maitland plant is a diversified operation producing chemicals, fibres, and plastics. At the time of the survey, it had about 700 people and an international union represented the workers. It also has an excellent safety record with 10 years without a lost work injury and only one medical treatment injury case in each of 1996, 1997, and 1998.

The plants are configured in a self-managing mode. Resource persons and others who perform duties previously done by supervision were included in the *supervision* category in the results. The results for corporate executives and for the management groups in the plants were not different enough to warrant separate treatment. They were combined as *management* in the database.

The combination of results from two quite different plants with those of the corporate executive group should provide a good picture of the company as a whole.

The highlights of the survey are discussed below. The questionnaire results are augmented with insights from the discussions, comments written in, and the interviews.

7.2.4   Insights from the DuPont Safety Survey

By a considerable margin, the DuPont statistical safety performance record is the best of the five very safe companies. In the results of the questionnaire survey, the DuPont answers were generally among the best, but not uniformly so. In many cases one of the other four companies had the "best result." This may be in part because of the high standards that have been in place for so long at DuPont. If safety is anything but exemplary, DuPont people tend to be self-critical. The managers at Whitby indicated dissatisfaction with their safety when they completed the questionnaire in 1996. The plant manager pointed out that although they had not had any lost work injuries for many years, their goal was to have no recordable injuries of any kind, and they had already had one in early 1996. They met this difficult goal of zero recordables in both 1997 and 1998.

7.2.4.1 Commitment of Management to Excellence in Safety. The commitment of DuPont leaders to safety is legendary. For generations, safety has had first priority. In DuPont Canada's "Directional Statement," safety is listed first:

> *Safety, Concern and Care for People, Protection of the Environment and Personal and Corporate Integrity* are this company's highest values and we will not compromise them.

For each of the values listed in the directional statement, beliefs, principles, and examples have been threshed out, communicated, and discussed throughout the company.

Safety is listed first in the objectives of every leader, of every person for that matter. Every management meeting starts with safety. At the time of this survey, the president led a discussion of safety as the first item when the executive team met every 2 weeks. In the infrequent case of a lost work injury, the responsible manager appears before this committee to explain the circumstances and discuss ways to prevent recurrence. In my experience, these were not situations of recrimination, because by DuPont's tenets the president and the executives were also accountable for the injury. DuPont requires that managers and supervisors, and their people, pay strict attention to safety. In that sense, there is an authoritarian aspect to it. The company has modified its approach as self-management has been introduced. Yet, in the end, the insistence of leadership that safety will be excellent is the driver. Generations of leaders have been steeped in this ethic. After a few years,

The Priority People Think Others Give to Safety					
	% Who Thought Others Rank Safety First				
	DuPont	Best Result	Safe Co. Avg.	Unsafe Co. Avg.	Worst Result
Managers' & Supervisors' View of Workers' Priority	72	96	81	39	41
Workers' View of Managers' & Supervisors' Priority	68	87	66	19	8

Figure 7-13   At DuPont Canada, workers and management have a balanced view of the priority given to safety (Q2).

these attitudes become built in to everyone's thinking. DuPont people everywhere are very proud of their safety.

In the priority that individuals say they give to safety, the DuPont results were at the average of the safe companies, all of which reported giving high priority to safety.

The perception of workers of the commitment of their leaders to safety and the perception of leaders of the commitment of workers was almost the same (Fig. 7-13). This balance is a healthy result, although the levels were just about at the average of the safe companies.

7.2.4.2   Safety Values. DuPont has long understood that the fundamental safety beliefs are the foundation of excellence in safety. The company established the beliefs decades ago and lives by them. Virtually 100% of DuPont respondents said that the company has clear, well-understood safety values. A high proportion felt that the values were up-to-date and influential (Fig. 7-14).

The DuPont responses to the questions that relate to specific values were also very positive. Safety was not seen to be in conflict with business objectives. Most respondents said that safety was well built in to the company operations.

More respondents in DuPont than in the other very safe companies subscribed to the belief that all injuries can be prevented (Fig. 7-15). This belief, important in itself, is also the basis for line management taking responsibility for prevention of all injuries.

7.2.4.3   The Extent to Which Management Is Held Accountable for Safety. In this important parameter, the DuPont results were the best

The Influence of Safety Values			
	% Who Said Safety Values Have ...		
	Important Influence	Some Influence	Little or No Impact
DuPont	79	17	4
Best Result	88	1	2
Safe Co. Avg.	67	22	11
Unsafe Co. Avg.	12	28	60
Worst Result	3	26	71

Figure 7-14 DuPont respondents consider safety values to be important (Q7).

The Belief That Injuries Are Preventable				
	% Who Believed *All* Injuries Can Be Prevented			
	Man.	Super.	Work.	All
DuPont	86	88	64	75
Best Result	86	88	64	75
Safe Co. Avg.	67	73	46	57
Unsafe Co. Avg.	25	23	15	20
Worst Result	0	0	11	9

Figure 7-15 The belief that all injuries can be prevented is held most strongly at DuPont (Q3).

of the very safe companies (Fig. 7-16). Even so, many DuPont leaders would be disappointed that more respondents did not say that management was *fully* responsible. It is one of the ironclad principles of the company that line management is totally responsible for the safety of its people and totally accountable for any injuries that do occur.

7.2.4.4 Involvement in Safety Activities. The level of involvement in safety activities reported by DuPont respondents was at the average of the results for the very safe companies (Fig. 7-17).

In the second part of Question 9, 44% of the DuPont respondents said that they had been on a safety committee, special task force, or team in the past 2 years. This was just at the average for the very safe

Line Management Accountability			
	% Who Said Line Management Is Held Accountable ...		
	Fully	**Fairly**	**Only Generally**	**Little Or Not At All**
DuPont	61	22	8	9
Best Result	61	22	8	9
Safe Co. Avg.	44	31	11	14
Unsafe Co. Avg.	10	6	47	37
Worst Result	4	6	60	30

Figure 7-16 Line management is held accountable for safety performance at DuPont (Q8).

Involvement in Safety Activities			
	% Who Said They Are Involved ...		
	Deeply	**Quite**	**Moderately**	**Not Much, Not At All**
DuPont	27	27	28	18
Best Result	46	28	18	8
Safe Co. Avg.	28	28	20	24
Unsafe Co. Avg.	8	21	21	50
Worst Result	3	13	7	77

Figure 7-17 Involvement is at a good level in DuPont (Q9).

companies. About 60% said that they were regularly involved in safety audits, one of the highest results in the survey.

7.2.4.5 Safety Training. DuPont respondents reported a somewhat lower level of safety training than the average of the very safe companies. In the plants, new employees get extensive training in safety. Even summer students get several days of such training before they begin work. Turnover is very low in DuPont plants, and most of the plant employees have many years of service. Perhaps that explains the results.

7.2.4.6 Safety Rules and Their Enforcement. DuPont respondents considered that their safety rules were of good quality and generally obeyed, answers at or near the top for both questions (Fig. 7-18). DuPont ranked with Milliken (the "best result") as thorough and consistent in enforcing safety rules.

7.2.4.7 Other Safety Practices. DuPont respondents generally gave a high rating to their safety practices—safety meetings, investigations, audit processes, etc.

DuPont respondents had no doubt about the importance of off-the-job safety (Fig. 7-19). The answers to this question were the highest by

Enforcement of Safety Rules			
% Who Said Disciplinary Action Taken For ...			
All Infractions	**Serious Infractions**	**Not Consistent Arbitrary**	**Seldom Taken**
DuPont	61	24	10	5
Best Result	67	20	12	1
Safe Co. Avg.	45	27	11	17
Unsafe Co. Avg.	18	21	27	34
Worst Result	4	16	11	69

Figure 7-18 The safety rules are well enforced at DuPont (Q14).

Off-the-Job Safety			
% Who Said Off-The-Job Safety Is ...			
Integral Part Of Program	**Present But Not Formal Part**	**Not Part Of Safety Program**
DuPont	99	0	1
Best Result	99	0	1
Safe Co Avg.	52	43	5
Unsafe Co. Avg.	10	35	55
Worst Result	7	7	86

Figure 7-19 Off-the-Job safety is stressed at DuPont (Q18).

Satisfaction with Safety Performance				
	% Who Said That They Are ...			
	Very Satisfied	Moderately Satisfied	Neutral	Dis-satisfied
DuPont	46	49	2	3
Highest Result	78	22	0	0
Safe Co. Avg.	53	43	2	2
Unsafe Co. Avg.	10	40	21	29
Lowest Result	10	10	27	53

Figure 7-20   Even though the safety performance is exceptional, people in DuPont are not totally satisfied (Q24).

far of the very safe companies. The company is unique in measuring off-the-job injury frequency.

Despite the company's excellent record, DuPont respondents were not totally satisfied with the safety performance (Fig. 7-20). They see room for improvement.

7.2.5   The Keys to DuPont's Continuing Excellence

DuPont's commitment to safety runs deep and goes back a long way. By the time a manager gets to a position of influence or a worker has a few years of experience, he or she has absorbed the strong company safety ethic. The commitment is thus robust and is sustained from generation to generation. DuPont has a very disciplined and persistent approach to safety. It is evidenced in a different way than at Milliken, for example, not so much in tight procedures as in embedded understanding of beliefs and principles. Safety pervades the culture of the company.

7.3   MILLIKEN AND COMPANY: WORLD CLASS SAFETY IN THE TEXTILE INDUSTRY

7.3.1   Introduction

Milliken is a unique company. Little known outside its own industry, it is one of the most successful companies in North America and perhaps the largest textile company in the world. Its approach to safety is as unique as its approach to business in general. In the past decade it has improved its record and is challenging DuPont for the label of the safest large company in North America.

Milliken started as the Deering Milliken fabrics company in the 1860s. Although it began in the northeast, it soon moved to join the emerging southern textile industry in South Carolina, which remains its central location. William Deering later left the company to found Deering Harvester, now known as Navistar. Milliken is a private company and has remained in the family for 130 years. Roger Milliken, the chairman, in his 80s, is still a driving force. Milliken is known for its obsession with customer service and quality. It was an early winner of the Baldridge quality award. It has long focused on research and maintains the world's largest textile research centre in South Carolina.

From its base in the southeast, Milliken has expanded worldwide. It has more than 70 manufacturing sites and about 16,000 people, largely in the US but also in Europe and Japan. It does not disclose its revenue, but it is known to be one of the world's largest suppliers of fabrics (and one of the world's largest privately held companies).

Milliken is committed to the "pursuit of excellence," seeking the latest in technology, in management systems, and in processes of all kinds. It is unusually committed to measuring everything and using the metrics for continuous improvement. It benchmarks against the best it can find. Once Milliken decides on a path, it has the focus and unity of purpose to implement it quickly, thoroughly, and throughout the company.

Milliken demands a great deal of its people, managers and workers alike. Yet it gives a lot back to them—great training, excellent working conditions, and the opportunity to participate in leading edge endeavours.

In the late 1980s, Milliken became committed to self-management. Typically it did not go at it with half measures but began to embed empowerment practices across the company. Today it is well along that path. In its plants, associates (workers) are involved in quality, continuous improvement, and safety management.

7.3.2 The Safety Record of Milliken

Milliken has always been a safe company compared with the textile industry as a whole and with US industry in general. In the 1980s, the company set out to become one of the world's safest companies. It is one of DuPont's largest customers, and DuPont is one of its largest sup-

The Ten-Year Safety Record of Milliken & Company											
	'89	'90	'91	'92	'93	'94	'95	'96	'97	'98	Avg. 1993-97
Milliken											
LWIF	0.27	0.08	0.21	0.14	0.13	0.10	0.05	0.08	0.09	0.05	0.09
TRIF	2.18	1.03	1.18	1.03	0.97	0.86	0.84	0.86	0.87	0.60	0.88
US Textile Industry											
LWIF											
TRIF						7.3	8.2	7.8	6.7		7.5

Figure 7-21 Milliken's safety record has improved continuously to world class.

pliers. For decades, the relationship has been one of teamwork, beneficial to both companies. Milliken set out to learn everything it could
about safety from DuPont but also sought the best practices wherever
they could be found. It views DuPont safety as the benchmark and has
set out to equal it. Today, some in Milliken say, "We have the best safety
systems and practices anywhere, at least equal to and possibly better
than DuPont's. In time we will have the best safety statistics."

Milliken's improvement in safety places it at the top of the US textile
sector and among the safest large companies in the world (Fig. 7-21).
Very few companies have a TRIF averaging less than 1.

7.3.3   The Safety Survey at Milliken

The safety survey at Milliken included these elements:

1. Examination of company safety data and other documents
2. Questionnaires completed by 103 employees at the Monarch and
   Gillespie, SC plants (total population of the two plants was
   350–400 people) and 27 people from the Corporate Safety
   Committee[vi]
3. At the Monarch plant:
   - Tour of the plant and discussions with managers and associates
   - Interviews with the plant manager and the manager of human
     resources
   - A focus group discussion with 15 workers (called associates)

[vi] The Committee, chaired by the Corporate Director of Safety and Health, included
people from across the company, mainly from southeastern plants and mainly managers but with some workers and professionals.

Responses to Questionnaires	
Group	No. of Questionnaires
Corporate Management	1
Other Managers	34
Total Management	35
Supervision	10
Professional	7
Working Level	77
Undesignated	--
TOTAL	129

Figure 7-22   Responses to questionnaires at Milliken.

- Discussion of the questionnaire results with a cross section of employees, including some of management group and the safety steering committee
4. Interviews and discussions with the Corporate Director of Safety and Health

Milliken's plant operations are configured in a self-managing mode, and few are called managers or supervisors. People who perform duties previously done by supervisors were included in *supervision* in Fig. 7-22. *Management* includes the managers at the plants and on the Corporate Safety Committee. Only one of the respondents, the Corporate Director of Safety, Wayne Punch, was included as a corporate executive.

The 130 questionnaires were completed carefully, and the interviews and focus groups were equally purposeful. Only one questionnaire was not useable.

The highlights of the survey are discussed below. The questionnaire results were augmented with insights from the focus groups and interviews. The results from the plants and the Corporate Safety Committee were merged. Inclusion of the corporate committee, mainly managers, perhaps distorted the results somewhat. However, the differences were not great enough to warrant separation.

7.3.4   Insights from the Milliken Safety Survey

Milliken's safety record is excellent—world class—and its processes are state of the art. Not surprisingly, the questionnaire results were among the best.

Milliken's approach to safety is similar to its approach to quality and business management—disciplined, orderly, with excellent systems meticulously implemented. The safety policies, systems, and procedures are the same in all of their plants, domestic or foreign. Milliken measures more things than do most companies, usually on a 4-week basis.[vii] They measure parameters that other companies consider "intangible" or not measurable. Significantly, these are the most important safety drivers. For example, workforce involvement, a "leading" indicator, is measured every 2 weeks worldwide. Practices are highly systematized. Once a practice has been established, it is thoroughly implemented across the company. Each plant has a safety audit once per year.

The company believes that safety is behaviour-based, and its policies and systems flow from that belief, hence the emphasis on involvement and on continuous training. The goal is to give everyone three things: education and training, healthy and safe processes, and the tools to do the job.

7.3.4.1 *Commitment of Management to Excellence in Safety.* Milliken has unusual confidence and determination to be the best in all that it does. Not surprisingly, this is evident in the approach to safety. There is strong commitment at all levels to eliminate injuries and to continue to improve, despite what might seem to be already a fine record. The company continuously seeks new initiatives to improve its safety. The commitment of management is evident in the words of Roger Milliken on the front of the brochure welcoming visitors to the company:

> The safety and health of all of its people is of primary importance to Milliken and Company. Milliken will devote resources to train its people to perform their jobs safely, to eliminate workplace hazards, and to comply with applicable safety and health laws and regulations. Milliken believes that all injuries are preventable, all health risks are controllable, and management is accountable.

The first question addressed to Tim Clever, plant manager of the Monarch plant, when he was interviewed in 1998, was "What are the three or four key beliefs that your company holds about safety?" The answer was unhesitatingly given:

[vii] Milliken follows the logical but unusual practice of keeping records, including financial data, on a 13 equal periods per year basis rather than the usual monthly accounting standard.

The Priority Individuals Give to Safety				
	% Who Ranked Safety First			
	Manag.	Super.	Work.	Avg.
Milliken	97	90	92	94
Best Result	97	90	92	94
Safe Co. Avg.	84	79	91	83
Unsafe Co. Avg.	46	64	65	62
Worst Result	44	58	53	56

Figure 7-23   Safety gets first priority at Milliken (Q1).

1. Safety has number 1 priority.
2. All injuries and incidents can be prevented.
3. Safety is not a cost but contributes to the bottom line.

Tim stresses safety in his contacts with people in the plant and in his frequent attendance at safety committee meetings. Safety is the first item at every management meeting. Once a week he has a lunch with a cross section of people, and safety is the first topic there.

The commitment of management, and of the workforce in general, showed up in the priority individuals give to safety (Fig. 7-23). In most of the very safe companies, individuals reported giving high priority to safety. The Milliken results were the highest.

The perception of workers of the commitment of their leaders is one of the best predictors of excellence in safety. Most of the Milliken workers said that their leaders put safety first, a very good result (Fig. 7-24).

7.3.4.2 *Safety Values.* Milliken calls safety and health its "Number One Corporate Value." Employees take a short refresher course on the Milliken safety values every year, pass a test, and sign off that they accept the safety values. As at DuPont, everyone at Milliken said that they have written safety values. Almost all of the Milliken respondents said that their values were up-to-date and influential, the highest result of the very safe companies (Fig. 7-25).

The Milliken responses to the questions on specific values were also very positive. Safety is not seen to be in conflict with business objectives. Most respondents said that improving safety even to exceptional levels would continue to yield benefits (Fig. 7-26).

The Priority People Think Others Give to Safety

	% Who Thought Others Rank Safety First				
	Milliken	Best Result	Safe Co. Avg.	Unsafe Co. Avg.	Worst Result
Managers' & Supervisors' View of Workers' Priority	76	96	81	39	41
Workers' View of Managers' & Supervisors' Priority	90	87	66	19	8

Figure 7-24 Milliken people are confident that others give high priority to safety (Q2).

The Influence of Safety Values

	% Who Said Safety Values Have ...		
	Important Influence	Some Influence	Little or No Impact
Milliken	88	10	2
Best Result	88	10	2
Safe Co. Avg.	67	22	11
Unsafe Co. Avg.	12	28	60
Worst Result	3	26	71

Figure 7-25 Safety values are very influential at Milliken (Q7).

The Cost-Benefit Break-Point

	% Who Said That at This Level Safety Starts To Cost More Than It Benefits ...			
	No Limit	Excellent	Good or Average	Always Net Cost
Milliken	92	7	1	0
Best Result	92	7	1	0
Best Co. Avg.	77	21	2	0
Unsafe Co. Avg.	67	12	12	9
Worst Result	47	23	20	10

Figure 7-26 There is little doubt at Milliken of the benefits of continuing to improve safety (Q5).

Line Management Accountability				
	% Who Said Line Management Is Held Accountable ...			
	Fully	**Fairly**	**Only Generally**	**Little Or Not At All**
Milliken	40	37	9	14
Best Result	61	22	8	9
Safe Co. Avg.	44	31	11	14
Unsafe Co. Avg.	10	6	47	37
Worst Result	4	6	60	30

Figure 7-27   Line Management is held accountable for safety performance at Milliken (Q8).

For a company already at an excellent level of safety, this is an important result. In some of the very safe companies, a significant number of people said that they believe that beyond achieving excellence in safety, further improvement would cost more than it yields in benefits. This is a barrier to further improvement in safety in these companies.

7.3.4.3   The Extent to which Management Held Accountable for Safety.   In this important parameter, the Milliken results were average for the very safe companies, below the best result (Fig. 7-27). This is surprising, given Milliken's disciplined approach, its safety record, and its outstanding results in most of the other questions.

It was not just workers who reported in this way; so did managers. To some extent this result may come from the characteristics of self-managing systems. When everyone takes appropriate responsibility for his or her own safety, the ultimate responsibility and accountability of line management is perhaps less evident. (However, self-management is also entrenched at some other benchmark companies.) In the interviews, managers said they felt totally responsible for the safety of their people. In the focus group of workers, there were some comments supporting less-than-perfect taking of responsibility. "There is room for improvement in managers' commitment." "Some managers need to be more supportive."

7.3.4.4   Involvement in Safety Activities.   Milliken respondents reported the highest level of involvement of the very safe companies (Fig. 7-28).

Involvement in Safety Activities				
	% Who Said They Are Involved ...			
	Deeply	**Quite**	**Moderately**	**Not Much, Not At All**
Milliken	46	28	18	8
Best Result	46	28	18	8
Safe Co. Avg.	28	28	20	24
Unsafe Co. Avg.	8	21	21	50
Worst Result	3	13	7	77

Figure 7-28   Involvement is high at Milliken (Q9).

In the second part of question 9, 74% of Milliken respondents said that they had been on a safety committee, special task force, or team in the past 2 years, the highest result from the very safe companies.

Milliken's goal is 100% involvement of everyone in safety activities beyond the specific job requirements—in committees, in audit teams, in conducting training, and in other such activities. Their measured involvement at the time of this survey was 68%. This corresponded well to the questionnaire results, where 74% that said they were deeply or quite involved and (somewhat coincidentally) where 74% that said that they had been involved in a special activity in the last 2 years. The correspondence between the results from the questionnaire and Milliken's data is noted in the discussion of question 9 in Chapter 6. Milliken respondents also said there was good involvement in audits, although not at the level of the highest of the safe companies.

In knowledge of safety goals and performance and in the extent of empowerment, Milliken answers also were among the highest of the very safe companies.

7.3.4.5 Safety Training. Milliken respondents reported the highest level of training of the very safe companies (Fig. 7-29).

Every new employee gets 40 hours of training, most of it on safety. Workers must take at least 32 hours of safety training every year and in most cases have considerably more. In the southeastern US, where most of Milliken's operations are located, the unemployment rate is very low and thus the turnover of workers is considerable. Milliken frequently hires people with little or no industrial experience. Consequently, the need for training is higher than at companies such as Shell

Safety Training of Workers					
	% of *Workers* Who Said Their Training In Last Two Years Has Been ...				
	Exten-sive	Consid-erable	Some	Little	None
Milliken	71	23	4	1	1
Best Result	71	23	4	1	1
Safe Co. Avg.	26	43	24	6	1
Unsafe Co. Avg.	1	7	29	25	38
Worst Result	0	4	8	29	59

Figure 7-29   The highest level of training in the survey was found at Milliken (Q11).

Enforcement of Safety Rules				
	% Who Said Disciplinary Action Taken for ...			
	All Infractions	Serious Infractions	Not Consistent Arbitrary	Seldom Taken
Milliken	67	20	12	1
Best Result	67	20	12	1
Safe Co. Avg.	45	27	11	17
Unsafe Co. Avg.	18	21	27	34
Worst Result	4	16	11	69

Figure 7-30   The safety rules are well enforced at Milliken (Q14).

and DuPont, where most of the workers have long service. Nonetheless the level of training at Milliken is undoubtedly one key to their excellent safety record.

7.3.4.6 Safety Rules and Their Enforcement. Milliken respondents said that their safety rules were of good quality and generally obeyed, answers at or near the top in both questions. Milliken reported a relatively high level of enforcement (DuPont was about the same), but even then well below the ideal (Fig. 7-30). Milliken has a zero tolerance policy for certain serious misdemeanors. For example, breaking the "lock, tag, and try" procedure calls for immediate dismissal.

Recognition for Safety Achievements			
% Who Said Recognition Is ...			
	Thorough Extensive	Frequent	Some	Little or None
Milliken	39	49	11	1
Best Result	92	8	0	0
Safe Co. Avg.	55	35	9	1
Unsafe Co. Avg.	7	15	27	51
Worst Result	7	7	14	72

Figure 7-31   Recognition for safety achievement was lower at Milliken than at some of the very safe companies (Q19).

7.3.4.7   *Other Safety Practices.* Milliken respondents generally gave a high rating to the quality of the safety practices. Their exemplary results for safety meeting frequency and attendance and their innovative "five levels of why" philosophy in investigations are highlighted in Chapter 6.

The perception of the extent of recognition for safety achievement varied quite widely, even among this group of high-performing companies. Milliken people did not rate recognition at as high a level as did respondents at some other companies (Fig. 7-31).

The Milliken respondents rated the safety of facilities and equipment, the safety organization, and the safety department relatively highly.

Despite their excellent safety record, they were less satisfied with their performance than some companies with poorer safety. This is a healthy result; they know that they can do better.

7.3.5   The Keys to Milliken's Continuing Excellence

Milliken is an unusual company with a unique culture and dedication to thoroughness and quality in all that it does. In meeting with and working with them, it became clear that they are truly world class in safety and that they would continue to improve. They have strong management commitment to safety. Their approach is characterized by thoroughness and discipline in installing the best practices and then continuously improving both the practices and the results. They are very thorough in measuring everything, even elements not quantified elsewhere. There is an authoritarian aspect in their insistence on meticulous practice and on achieving results. Yet they have very deep participation by their workers. They will give DuPont a run for its money as the safest large industrial company in North America.

7.4 S&C ELECTRIC CANADA: A TURNAROUND TO SAFETY EXCELLENCE IN THE ELECTRICAL EQUIPMENT INDUSTRY

7.4.1 Introduction

In the late 1980s, S&C Electric Canada had a poor safety record, with more than 20 lost work injuries among 250 employees in each of 1988 and 1989. Then in a complete turnaround, they had only one lost work injury in 1992! For 5 years from mid-1992 through mid-1997, they had *no* lost work injuries. The total injury frequency was reduced 10-fold.

S&C Electric Canada is a subsidiary of an American company with an average safety record. It appears that S&C Canada has achieved excellence in safety through its own local initiative. How did they do it? What can be learned from their experience?

The S&C Electric Company is fairly small, with fewer than 2000 people, mainly in Chicago and Toronto. It is a world leader in high-voltage switching and fuses. The company's unique character derives from its history and its family ownership. It was started in 1911 by two electrical engineers, Edmund Schwietzer and Nicholas Conrad (thus S&C), who saw the need for a special fuse. The Conrad family still owns S&C, and John Conrad, son of the co-founder, is still active. The company has a reputation for quality, reliability, and integrity. Their main customers are utilities, but they also sell to other industries worldwide. Their strategy is based on growth through research.

The Canadian subsidiary was started in 1953. The Toronto plant was built in 1962 and has expanded several times. The Canadian Company produces some lines for the whole company, exporting to the US and offshore. The president of the US company, John Estey, a Canadian, is also chairman of S&C Canada.

The Canadian head office is located in the Toronto plant. The plant is meticulously clean and orderly, unusually so for an operation involving metal fabrication. The in-process inventory is right on the factory floor, and, with quality processes, there is not much of it. The machinery is modern and automated. Still, there are many of the hazards typical of fabrication and assembly operations.

The Ten-Year Safety Record of S&C Electric											
	'89	'90	'91	'92	'93	'94	'95	'96	'97	'98	Avg. 1993-97
S&C Electric											
LW Injuries #	24	12	11	1	0	0	0	0	1	1	~
LWIF	8.2	4.2	3.9	0.4	0	0	0	0	0.31	0.29	0.06
TRIF	13.2	9.7	6.4	1.2	2.0	2.0	2.4	2.0	2.5	2.0	2.2
Elec. Rate Group.											
LWIF, Avg.	7.0	6.7	5.9	5.2	3.1	2.2	1.6	1.8	1.3	1.4	2.0
The Electrical Rate Group is for like manufacturers in the Province of Ontario											

Figure 7-32   The remarkable turnaround in safety performance at S&C Canada.

Seldom have I visited a company so friendly, where everyone says hello. The company pays well, but not unusually so, and has good benefits and a profit-sharing plan for all employees. They have a no-layoff strategy, and turnover is low. S&C Canada has made progress in creating a participative work environment, eliminating differences between the treatment of workers and staff. The operation is non-union. The atmosphere of trust undoubtedly has much to do with the excellence in safety and vice versa.

7.4.2   The Safety Record of S&C Electric Canada

The safety performance of S&C Electric is given in Fig. 7-32. The LWIF record is very good, but the total injury record is not world class. The safest companies have lower frequencies, some with TRIFs well below 1.0. Nevertheless, S&C stands out as an example of excellence in safety, particularly in the turnaround in the 1990s.

Many of the very safe companies in Canada are subsidiaries of foreign companies with excellent safety. S&C Canada's parent had an average safety record until its recent substantial improvement (Fig. 7-33).

7.4.3   The Story of the 1989–1993 Turnaround

In the late 1980s, Bill Lowry, a long-time employee, was appointed president of S&C Canada.[viii] Lowry, an engineer, had spent much of his life in sales. He was committed to participative management and brought strong values about people to his new job. S&C was aware of the costs

---

[viii] In 1997, Lowry retired and Grant Buchanan, previously VP of Engineering Services, became president.

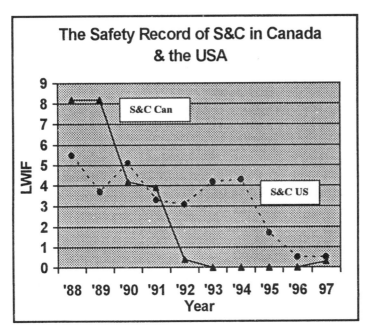

Figure 7-33 Lost work injury frequency has improved at both the parent company and the subsidiary.

of their bad safety—costs in compensation, in lost production, and in other less tangible ways. Lowry saw that changes in management style and in employee attitude were needed to improve not only safety but also quality and costs.

Under Lowry's leadership, S&C began to change to a more open, participative culture. This involved more than safety, but safety was a key element. Bill Lowry, Peter Pillon, now Director of Production, and Doug Patten, now Director of Human Resources, were on the team that drove the change. Doug was the safety supervisor. They were all on the CMTS team ("Corporate Minding the Store") at the time of this research survey.

Establishing a vision, values, and goals was the first step. In 1989, S&C set a goal of cutting lost work injuries in half, from 24 in 1989 to 12 in 1990. If they reached the goal, an employee would win a trip for two to Las Vegas in a draw. They met the goal. The early achievements were widely celebrated, and recognition became an important part of reinforcing their progress. When an injury terminated their record at 88 days without a lost work injury, Bill Lowry remembers thinking "We'll never get to that level again!" But success built on success. The

1990 achievement gave them confidence. The breakthrough came in 1992, when they had just one lost work injury. They went on through a 5-year period without a lost work injury.

The leaders at S&C say that establishing safety as first priority was the most important step. Changing the attitude of management and then getting the workforce to believe that the commitment was real was difficult. Lowry plays down his role, emphasizing that it was a team effort. However, in an environment with a strong production culture, he convinced managers that safety was not in conflict with production but instead would improve effectiveness. In his retirement speech, Bill Lowry said that his best memory was the achievement in safety. He cited commitment of management as the key. "Getting people to believe that we believed in safety was the hardest thing."

The change to a participative culture was another important step in the turnaround. A continuous improvement program involving all employees was initiated. Supervisors were trained to be more participative. Employees saw that their input was encouraged and that good suggestions would be implemented. Training was increased. The change in attitude was also essential in improving quality, costs, and productivity. The physical aspects of safety were not neglected. There was much greater effort to automate and to apply ergonomics to equipment and processes. People have real pride in their jobs and in the workplace. It can be seen in the excellent housekeeping.

The comments written in on the questionnaires by workers are evidence of the success in creating a positive environment for safety. Here are some of them:

- The company states that quality and safety are our number one concern. I think that they show it. The president is really involved with the safety program. The company cares.
- My job allows me to visit a great number of factories. With my heightened safety awareness, I find myself shaking my head in disbelief at what is acceptable in these plants. The hardest part of a safety inspection at S&C is to actually find a safety infraction.
- The biggest change in attitude in our company in the last 6 years has been at the blue-collar level. People working directly in manufacturing have become more self-aware at working safely.
- S&C is a good and safe place to work. There is a strong commitment to keep the facility clean and safe. I am very happy that we have introduced the concept of "teamwork" (i.e., doing many different jobs so as not to overstress certain muscles and avoid boredom).

· Safety is an attitude. S&C has changed theirs, and most of the employees have too. Our safety program is not perfect, but we are trying hard. One day at a time. If you looked at S&C 8–10 years ago, you would not know it today. The company cares for its employees. It is a great place to work. Eight to ten years ago it was only a job.

At a dinner of the quarter-century club, John Conrad noted that the safety record of the Canadian operation was better than in the US and asked, "How is it happening?" An employee indicated how things had changed. "It's really just attitude. Five or ten years ago you would see a guy on the top rung of a ladder and people would be taking bets on how soon the sucker would fall. Today there would be six guys shouting at him."

7.4.4 The Safety Survey at S&C Electric Canada

The safety survey at S&C Electric Canada included these elements:

· Examination of company safety data and other documents
· Safety questionnaires completed by 65 employees
· Interviews with the president and the directors of production and human resources
· Two focus groups—one of 12 supervisory people and one of 12 working level people

Sixty-five of the three hundred people in the company completed the questionnaire, including the president, the CMTS team, other managers, supervisors, professional staff, and workers (Fig. 7-34).

The questionnaires were completed carefully and returned promptly. The focus groups were equally purposeful. People were very cooperative, seeing the research as an opportunity to learn and improve as well as recognition for their excellent safety.

7.4.5 Insights from the S&C Safety Survey

The answers to the questionnaire were in line with the excellent safety at S&C. Some were the highest of the safe companies, even though the safety record of S&C is technically not at the level of some of the other companies (in TRIF). This may be caused by the great improvement and to the excellent environment of teamwork.

Highlights include the commitment to safety, the consistency in fol-

Responses to Questionnaires	
Group	No. of Questionnaires
Corporate Management	5
Other Managers	6
Total Management	11
Supervision	13
Professional	11
Working Level	30
Undesignated	-
TOTAL	65

Figure 7-34 Responses to questionnaires at S&C Electric.

lowing safety rules, and the quality of safety meetings and the modified duty and return-to-work systems.

7.4.5.1 Commitment of Management to Excellence in Safety.

The commitment to safety is evident in the words and actions of company leaders. The S&C Canada vision says: "We will accomplish this (their vision of a high-technology company) while retaining our enduring principles of integrity, fairness, honesty, and respect for the safety, well-being, and advancement of all S&Cers." The statement emphasizes programs to ensure the health and safety of employees, provision of a safe, modern, well-maintained facility, and the use of ergonomic principles to ensure safety and improve the comfort of employees. The Safety Policy Committee, comprising the senior leaders and chaired by the president, meets at least quarterly and prides itself on quick action to resolve issues. Once the problem was identified, a decision was made in less than a week to invest in a safer truck unloading system. After each meeting, the committee does a plant audit.

In the priority that respondents say that *others* give to safety, the results were the best of the safe companies (Fig. 7-35). S&C workers have a high opinion of the commitment of their managers and supervisors. Managers and supervisors have an equally high opinion of the workers' commitment. The level of the answers and the agreement between workers and leaders were higher at S&C than at the other very safe companies.

The perception of workers of the commitment of their leaders to safety is one of the best indicators of excellence.

The Priority People Think Others Give to Safety					
	% Who Thought Others Rank Safety First				
	S&C	Best Result	Safe Co. Avg.	Unsafe Co. Avg.	Worst Result
Managers' & Supervisors' View of Workers' Priority	96	96	81	39	41
Workers' View of Managers' & Supervisors' Priority	87	87	66	19	8

Figure 7-35   The S&C responses to this important question were the best found in the survey (Q2).

7.4.5.2 Safety Values. S&C people believe that their values provide a framework for excellence. They say that safety has first priority. They also say "safety, quality, and serving the customer are all number one." In many companies this would mean trouble, because safety does not have the same constant pressure as costs, quality, and volume and it might be pushed down in priority. Not at S&C. People there understand that safety and business performance must both be excellent but that in case of conflict, safety takes priority. There is a strong belief that all injuries can be prevented. The respondents said that safety is well built in to all that they do. Workers in the focus group were sure that they shared the responsibility to see that no shortcuts were taken. They pointed out that doing the job safely was part of "doing it right the first time" and would save effort in the long run.

7.4.5.3 The Extent to Which Line Management Held Accountable for Safety. The respondents did not rate S&C as well as some companies in the assessment of the responsibility line management takes for safety (Fig. 7-36). There was a perception, particularly among workers, that line management was not held fully accountable.

S&C management found this result hard to understand. Partly, it may be a problem of terminology. S&C has stressed that each person is fully responsible for his or her own safety. Unless the "internal responsibility system" is well understood, this might lead to the belief that supervision and management are not also responsible. Comments written in and raised in discussion groups indicate that this might be the case.

Line Management Accountability				
	% Who Said Line Management Is Held Accountable ...			
	Fully	Fairly	Only Generally	Little Or Not At All
S&C Electric	37	31	13	19
Best Result	61	22	8	9
Safe Co. Avg.	44	31	11	14
Unsafe Co. Avg.	10	6	47	37
Worst Result	4	6	60	30

Figure 7-36 Management accountability was rated good at S&C Electric but not as high as at some other safe companies (Q8).

Involvement in Safety Activities				
	% Who Said They Are Involved ...			
	Deeply	Quite	Moderately	Not Much, Not At All
S&C Electric	17	24	18	41
Best Result	46	28	18	8
Safe Co. Avg.	28	28	20	24
Unsafe Co. Avg.	8	21	21	50
Worst Result	3	13	7	77

Figure 7-37 Involvement was not rated at as high a level at S&C as at other safe companies (Q9).

7.4.5.4 Involvement in Safety Activities.

In Fig. 7-37, involvement at S&C is compared with that at other companies.

The average of the very safe companies was not as high as might be expected. Even so, the S&C results lag. The focus groups confirmed that formal involvement is not stressed but that there is involvement through work teams. The JH&S committee, called the Safety Practices Committee, is active, but it includes only 12 people.

Perception of involvement is subjective, so respondents were also asked whether they had been on a committee, task force, or team in the previous 2 years. Two-thirds (70% of workers) said they had *not*, lower than in most of the other very safe companies.

Involvement of *Workers* in Safety Audits			
	% of *Workers* Who Said That They Are ...		
	Regularly Involved	Have Some Involvement	Not Involved At All
S&C Electric	23	20	57
Best Result	68	27	5
Safe Co. Avg.	40	31	29
Unsafe Co Avg.	7	10	83
Worst Result	7	10	83

Figure 7-38   Involvement of workers in safety audits was fairly low at S&C Electric (Q16).

Most safe companies involve their people in audits of the workplace. A minority at S&C said they were involved in such audits (Fig. 7-38). However, S&C respondents reported that the audits that are done are of good quality. Many of those involved in audits were probably on one of the two standing committees.

7.4.5.5   Safety Rules and Their Enforcement. The S&C respondents judged their safety rules to be of good quality. They had the highest perception of the very safe companies on the extent to which rules are obeyed (Fig. 7-39). As with people at all companies, they were reluctant to say that there were *no* exceptions.

S&C has a written disciplinary policy. Although it does not make a big thing of the use of disciplinary action, safety is not to be taken lightly. The view of the Production Manager, Peter Pillon, is that "if people know what is expected of them and if it doesn't materialize, they'll be asked why. We don't keep quiet. We try to be positive but sometimes discipline is necessary."

7.4.5.6   Other Safety Practices. S&C respondents reported a fairly high level of training. One feature is quarterly training sessions for all employees. The company stresses continuous improvement: "What we do today will not be good enough for tomorrow." Although some training is done by the safety coordinator, supervisors have direct responsibility for training their people, and much of the training is coaching on the job.

Workers' Views on the Observance of Safety Rules			
	% Of Workers Who Said That Rules Are Followed ...		
	Without Exception	Generally	Sometimes	Often Not, Little Attention
S&C Electric	37	60	0	3
Best Result	37	60	0	3
Safe Co. Avg.	25	64	10	1
Unsafe Co. Avg.	4	35	54	7
Worst Result	0	27	64	9

Figure 7-39 Workers at S&C Electric reported that the safety rules are obeyed (Q13).

Almost all injuries and incidents are investigated, and action is taken—one of the best results of the companies surveyed. Peter Pillon confirms this. "Every close call, every medical treatment injury is investigated, to identify the obvious cause and to seek out the underlying cause."

Recognition—"tokens of a job well done"—has helped keep safety in front of everyone. S&C establishes "mini-milestones." If they reach them, there is a celebration—a BBQ, a rally, or some other visible way to reinforce the goal of eliminating injuries.

The safety of the facilities received the highest marks among the companies surveyed. A good rating was given to their modified duty and return-to-work systems (Fig. 7-40).

S&C respondents were satisfied with their safety organization and safety department, giving them among the highest ratings of the very safe companies. At the time of the survey, the safety department consisted of a safety coordinator with assistance from the Director of Employee Relations. They seem to have found a balance of involvement and coaching without taking away from the line ownership principle. S&C respondents reported a high level of satisfaction with their safety performance—not surprising, given the dramatic improvement.

Although S&C people gave their safety meetings good marks for quality, they were not as frequently held nor as well attended as in some of the other companies.

Modified Duty & Return-to-Work Systems			
	% Who Rated Modified Duty & Return-to-Work Systems As …		
	Excellent	Good	Satisfactory	Poor or Very Poor
S&C Electric	70	28	2	0
Best Result	70	28	2	0
Safe Co. Avg.	53	29	11	7
Unsafe Co. Avg.	13	20	23	44
Worst Result	13	20	23	44

Figure 7-40   S&C Electric people gave high marks to this important practice (Q17).

Off-the-job safety is not stressed as much as in some other very safe companies. S&C has realized this and has initiated new programs under a "Safety For Life" umbrella.

7.4.6   The Keys to the Turnaround to Safety Excellence

Management at S&C Canada is highly committed to excellence in safety. They "walk the talk." This is the "Essential Cornerstone" in the model of safety management. There is a unified commitment to safety among all employees. Workers feel a sense of ownership. They are confident that they can and will prevent injuries.

The environment of trust and employee ownership of safety is notable. Unlike in many such surveys, there were no negative comments. The same strong feeling of team ownership among all employees was evident in the group discussions. Seldom have I conducted a focus group of such positive working-level employees. They could not be drawn out to make any real criticism of the way that safety is handled at S&C.

7.5   SHELL CANADA: WORLD CLASS SAFETY IN THE OIL INDUSTRY

7.5.1   Introduction

The large oil companies have better safety records than industry in general. Perhaps, like the chemical industry, this results from the potential for major disaster. Among oil companies worldwide, Shell stands out as one of the safest, if not the safest, and Shell Canada has one of the best records in the Royal Dutch/Shell Group. Through the 1990s, Shell Canada's safety record improved continuously. With LWIF below 0.1 and TRIF below 0.8, it is one of the safest large companies in Canada. It is also one of the more profitable companies in its peer group. Shell Canada was thus a natural subject for research on safety management.

Shell Canada has operated in Canada for many decades and is one of the largest integrated oil companies in the country. It is a subsidiary of the Royal Dutch/Shell Group, headquartered in The Netherlands and the UK, but operates relatively independently, with about one-quarter of its stock on Canadian exchanges, with a full spectrum of corporate functions, and with largely Canadian management. Its scope is diverse, involving exploration, production, and refining operations in conventional oil and gas, heavy oil, tar sands, and offshore oil and gas. It has refineries in Quebec, Ontario, and Alberta and a national network of service stations. At the time of this research, it had about 3500 people in total. It had recently sold its chemicals operations, some to Shell Chemical International.

7.5.2 The Safety Record of Shell Canada

Shell's steady improvement in safety is shown in Fig. 7-41. In 1997 there were three lost work injuries in a workforce of about 3500; 1998 was a still better year, setting new records. Very few companies have a TRIF averaging less than 1. Shell's record is obviously in the top rank — world

The Ten-Year Safety Record of Shell Canada											
	'89	'90	'91	'92	'93	'94	'95	'96	'97	'98	Avg. 1993-97
Shell Canada											
LWIF	0.70	0.22	0.25	0.15	0.10	0.18	0.17	0.08	0.08	0.03	0.12
TRIF	1.62	0.75	1.10	0.87	0.79	0.83	0.50	0.59	0.55	0.40	0.65
CPPI											
LWIF					0.65	0.75	0.54	0.48			~ 0.6
TRIF					1.74	1.73	1.69	1.60	2.5		1.85
Notes: Industry average figures, provided by the Canadian Petroleum Products Institute (CPPI), include data for their member companies in oil and gas refining and marketing.											

Figure 7-41 Shell Canada's fine record of continuous improvement to world class.

class safety. Like many of the safest Canadian companies, Shell is the subsidiary of a foreign parent with a history of excellence in safety. In the 1990s, Shell Canada's performance has been close to or slightly better than that of Shell US and of the other companies in the international Shell group.

Among the large integrated oil companies in Canada, Shell has the best safety record, although Imperial Oil is not far behind. Although Shell Canada's record is better than the average of the Canadian Petroleum Products Institute (CPPI) member companies, the comparison to a relatively narrow slice of the North American industry does not do justice to its fine performance. The CPPI average is dominated by the performance of a few large, safe companies such as Shell and Imperial Oil. Shell's average TRIF of 0.65 compares with the CPPI average of about 1.9 and the average for comparable US companies of over 4 (American Petroleum Institute member companies).

Throughout its worldwide operations, Shell stresses contractor safety. Although the direct responsibility lies with the contractors, Shell's goal is for them to be as safe as Shell employees are. Contractor safety has improved, now averaging below 1.0 in LWIF, but it is not as good as that of employees, reflecting the difficulty in managing safety when contractors are in Shell facilities for short periods on shutdowns or for special functions.

Within Shell, the Products Group (refining and until recently chemicals) has steadily improved its safety. The Resources Group went through a period in the early to mid-1990s where their safety record deteriorated. They worked very hard to get back on track and in 1997 and 1998 had no lost work injuries. The TRIF of Resources has also improved to well below 1.0.

7.5.3   The Safety Survey at Shell Canada

The safety survey at Shell Canada included these elements:

1. Examination of company safety data and other documents
2. Questionnaires completed by 49 people at the Scotford, Alberta refinery (total population about 220) and by the CEO and 6 other corporate officers
3. Research at the Scotford, Alberta, refinery:
   - Questionnaires completed by 49 employees
   - Tour of refinery and discussions with the workforce

- Interviews with the General Manager, Manufacturing, Western Canada (who is also the refinery manager); the Manager of Operations for the refinery, and the Manager, Health, Safety and Environment
- Focus group discussions with a dozen supervisory people (called coordinators) and with a dozen working-level people
- Discussion of the questionnaire results with a cross section of refinery employees, including most of the senior management group

4. Research at the Corporate Executive level
   - Questionnaires completed by the CEO and his team (7)
   - Interviews with the CEO, the Senior Operating Officer, Products, and the Director, Health Safety and the Environment

Fifty-seven people completed the questionnaire of the 65 that were sent out, and one questionnaire was not useable (Fig. 7-42). The survey included all of the corporate executive group (7) and a cross section of people in the Scotford refinery.

The refinery operates in a self-managing mode. Coordinators, who perform some of the duties previously done by first-line supervision, are included in the *supervision* category in the questionnaire survey results. The corporate executives tended to rate the beliefs and practices of the company somewhat higher than did the refinery managers, but for most of the questions the differences were not great. The results for both are included together as *management* in the data summaries.

Surveying one-quarter of the workforce should give a statistically valid picture of the safety beliefs and practices at the refinery. The refin-

Responses to Questionnaires	
Group	**No. of Questionnaires**
Corporate Management	7
Other Managers	7
Total Management	14
Supervision	15
Professional	7
Working Level	20
Undesignated	--
TOTAL	56

Figure 7-42   Responses to questionnaires at Shell.

ery results combined with those of the corporate executive group should provide a reasonable picture of the company. However, a broader survey would be needed to be totally valid statistically. (The Scotford refinery safety record is somewhat better than in some other Shell operations.)[ix]

The questionnaires were completed carefully and returned promptly. The focus groups were equally purposeful. Overall, the company and its people were very cooperative, seeing the research as an opportunity for recognition for their excellence in safety as well as a chance to learn and improve further.

The highlights of the safety survey are discussed below. The questionnaire results are augmented with insights from the focus groups and the interviews.

7.5.4   Insights from the Shell Safety Survey

Shell Canada's safety record is consistently excellent—world class. In the group of five very safe companies, it is about at the median and the survey results were generally in the middle of the results for the five companies.

To achieve its excellent level of safety performance, Shell Canada does most things right. However, the safety survey and the supporting interviews and focus groups indicated some areas where Shell Canada could improve. Individuals say they give high priority to safety, but many did not think that management ranks safety as highly. There was a perception of a limit to the benefits of excellence in safety and a feeling that perhaps Shell was at the level of "diminishing returns." Although the safety rules were judged to be of good quality and generally followed, they were not as consistently enforced as in some companies. Involvement, particularly in audits, was lower than might be expected in a self-managing environment.

7.5.4.1   Commitment of Management to Excellence in Safety. There

is a high level of commitment to eliminating injuries at all levels of the company and a determination to continue to improve, despite what might seem to many to already be an excellent enough record. Shell sets its sights high. The president, Chuck Wilson, and other leaders

---

[ix] In 1996, a similar survey was completed by a cross section of Shell Resources employees as part of a consulting contract with the author. The objective was to assist Resources in its efforts to improve safety.

emphasized that being the best in the oil industry is not good enough. Shell's goal is to equal the best in industry worldwide. They strive to improve the safety of contractors, setting the same standards for them as for their own employees. If part of the company lags, as Resources did in the early 1990s, Shell exerts a strong effort to get it back on track. The company is always seeking new ways to improve safety. It has found ways to continue excellence in safety under the lean management and worker empowerment of self-management. It has good safety systems, structures, and processes.

The commitment of management to excellence in safety is evident in the words and actions of company leaders. CEO Chuck Wilson put it this way:

> It's management attention. Focus. You have to have the commitment. The commitment starts at the top but it has to be worked right down through the management chain. You have to have the people out there at the "coal face" who are involved and believe that we are serious about it — that we mean what we say when we say that safety is number 1.

Shell believes that unless safety and environmental protection are at high levels, the company cannot consider itself successful, despite how well other things might be going. The consistency with which this philosophy was repeated at various levels and locations shows how well this philosophy has been entrenched.

The commitment to safety is exhibited in the practice of reviewing safety first on the agenda at every significant meeting, from daily plant meetings on the shop floor to every meeting of the board of directors. Four times a year, the top 50 people in the company meet to review corporate safety.

The commitment of Shell also showed up in the priority that individuals reported that they give to safety (Fig. 7-43). Most respondents reported giving high priority to safety.

Although Shell respondents *as individuals* said that they give high priority to safety, they did not think that others in their organization give safety as high a priority (Fig. 7-44).

Managers and supervisors believed that workers are committed. Workers, however, had some ambivalence about the commitment of managers and supervisors. There was more divergence than in most of the very safe companies. In the poorly performing companies, safety is not given high priority and the divergence between the perception of workers and that of managers is pronounced.

The Priority Individuals Give to Safety				
	% Who Ranked Safety First			
	Manag.	**Super.**	**Work.**	**Avg.**
Shell	92	84	95	87
Best Result	97	90	92	94
Safe Co. Avg.	84	79	91	83
Unsafe Co. Avg.	46	64	65	62
Worst Result	44	58	53	56

Figure 7-43   Shell respondents said that as individuals, they give high priority to safety (Q1).

The Priority People Think Others Give to Safety					
	% Who Thought Others Rank Safety First				
	Shell	**Best Result**	**Safe Co. Avg.**	**Unsafe Co. Avg.**	**Worst Result**
Managers' & Supervisors' View of Workers' Priority	86	96	81	39	41
Workers' View of Managers' & Supervisors' Priority	53	87	66	19	8

Figure 7-44   Workers' view of management's priority was lower at Shell than at some other safe companies (Q2).

7.5.4.2 Safety Values. Almost all Shell people (92%) said that the company has safety values. However, they did not attribute as high an influence to them as respondents in some other companies.

Shell people were also less certain that they could continue to improve beyond their already fine record. A larger proportion of respondents than in the other very safe companies felt that beyond achieving excellence in safety, further improvement might cost more than it would yield in benefits (Fig. 7-45).

7.5.4.3 The Extent to Which Line Management Held Accountable for Safety. For this parameter, the Shell results were at the average of the very safe companies.

The Cost-Benefit Break-Point				
	% Who Said That at This Level Safety Starts To Cost More Than It Benefits ...			
	No Limit	Excellent	Good or Average	Always Net Cost
Shell	53	43	4	0
Best Result	92	7	1	0
Best Co. Avg.	77	21	2	0
Unsafe Co. Avg.	67	12	12	9
Worst Result	47	23	20	10

Figure 7-45   Shell respondents were less certain than others that improvement beyond excellent safety would be beneficial (Q5).

Involvement in Safety Activities				
	% Who Said They Are Involved ...			
	Deeply	Quite	Moderately	Not Much, Not At All
Shell	16	32	20	32
Best Result	46	28	18	8
Safe Co. Avg.	28	28	20	24
Unsafe Co. Avg.	8	21	21	50
Worst Result	3	13	7	77

Figure 7-46   Involvement was somewhat lower at Shell than at the other safe companies (Q9).

7.5.4.4   Involvement in Safety Activities.   Shell respondents reported somewhat lower involvement in safety activities than the average of the safe companies (Fig. 7-46). *Workers* reported a lower involvement than managers or supervisors. The same pattern was observed in the question about specific involvement in committees, task forces, or special safety teams. Two-thirds of the Shell respondents (80% of workers) said they had not been so involved in the previous 2 years.

In knowledge of safety goals and performance, and in the extent of empowerment, Shell answers were at about the average of the very safe company results.

7.5.4.5 Safety Training. The Shell responses indicated a lower level of training than at some of the other very safe companies, perhaps as a result of their very stable, long-service workforce.

7.5.4.6 Safety Rules and Their Enforcement. The Shell respondents judged their safety rules to be of good quality and generally obeyed. As with people at all companies, respondents were reluctant to say that there were *no* exceptions.

As with most companies, even some of the safer ones, Shell was not perceived to consistently enforce the observance of safety rules (Fig. 7-47). In this respect, the answers were at about the average for the very safe companies.

7.5.4.7 Other Safety Practices. The perception of the extent of recognition for safety achievement varied quite widely, even among the group of very safe companies. Shell people did not feel that recognition was at as high a level as did the respondents at some other companies.

Enforcement of Safety Rules			
% Who Said Disciplinary Action Taken For ...			
All Infractions	Serious Infractions	Not Consistent Arbitrary	Seldom Taken
Shell	33	38	13	16
Best Result	67	20	12	1
Safe Co. Avg.	45	27	11	17
Unsafe Co. Avg.	18	21	27	34
Worst Result	4	16	11	69

Figure 7-47 Enforcement of safety rules at Shell was somewhat lower than the average of the safe companies (Q14).

7.5.5   The Keys to Shell's Continuing Excellence.

In meeting with and working with Shell at various locations and levels of the company, one gets the strong impression that they are in it for the long term and that there is an enduring drive for excellence. The standards are high, and continuous improvement is persistently sought. Like DuPont, Shell has a long history of working to improve safety and its employees tend to be critical if things are less than ideal. This showed up in the survey results. It is significant that Shell is not content with its excellent level of safety, not content to just be leaders in the oil industry; instead it seeks to match the "best of the best." When the Resources Division faltered, a long-term effort was mounted to get it back on track. High value for safety is indeed built into the culture.

8

CONCLUSIONS–HOW COMPANIES ACHIEVE EXCELLENCE IN SAFETY

The model of safety management (Fig. 8-1) was used as a framework for discussion of the results of the research.

The key elements of the model are:

- **The commitment of management** to excellence in safety
- **Line management ownership** of the safety agenda
- **Involvement in safety activities**, complemented by training
- **Comprehensive safety practices**
- **The safety organization and safety specialists**

THE COMMITMENT OF MANAGEMENT TO EXCELLENCE IN SAFETY

In all aspects of this research, the strong commitment of the leaders of the five very safe companies stands out, confirming that it is "the essential cornerstone" of excellence in safety. The commitment was observed time after time. Management commitment was evidenced in the questionnaire results and in the observations, interviews, and group discussions in the very safe companies. The DuPont practice of the president leading a safety discussion as the first item in the biweekly management meeting was a typical indication of the depth of commitment. The commitment of management to safety at S&C Electric was captured in the comment written in on the questionnaire by one worker: "The

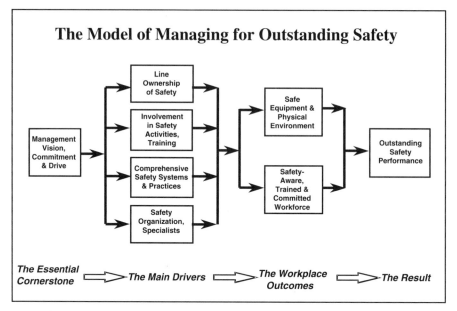

Figure 8-1   The model identifies the main factors in safety management.

company states that quality and safety are our number one concern. I think that they show it. The president is really involved with the safety program. The company cares."

Support for the importance of management commitment was evident in the questionnaire results. In the first question, 84% of the managers in the very safe companies said that they give first priority to safety, compared with 46% in the companies with poor safety. Even more importantly, the workers in the safe companies believed that their management puts safety first: 68% said their managers give first priority to safety. In the companies with poor safety, managers did not claim to give priority to safety and their workers were positive that they do not. Only 18% of the workers polled said that their managers gave safety first priority. In neither of these questions was there any overlap in the answers—the worst answer from the very safe companies was higher than the best from the companies with poor safety.

The leaders of the very safe companies base their safety management on well-thought-out and thoroughly communicated safety values. Not surprisingly, this showed up in the answers to question 7 about values. In the safe companies, 90% of all respondents said their organization has written safety values and 67% said the values were up-to-date and influential. In the companies with very poor safety, 50% thought there were values but only 12% said they were influential. This contrast was clearly evident in the answers to other questions dealing with values, for example, in question 3 (the extent to which injuries are preventable), in questions 4 and 5 (the interaction between safety and business), and in question 6 (the extent to which safety is built in).

The great contrast between the beliefs and the practices of managers in the very safe companies and those in the companies with poor safety emerged in almost all of the results.

Management commitment is undoubtedly the foundation of safety. Without it, the rest of the agenda for excellence can not be effective. And it must be real, sustained, determined, and believable. It means that the leaders understand safety, believe in it with passion, and see that their passion is embedded in the company's culture. They see that all managers have clear, make-happen safety objectives that are audited regularly. They insist on high standards, on careful measurement, on benchmarking against the best. If world class safety performance is not forthcoming, they will make changes and keep at it until it is.

LINE MANAGEMENT OWNERSHIP OF THE SAFETY AGENDA

Line ownership of the safety agenda is a direct extension of management commitment. Managers, supervisors, and workers understand and accept the organization's values and goals and convert them into practice. In the very safe companies, this shows up in many ways. All managers have specific safety objectives and they must get results. The comment of Sam Spanglet, refinery manager at Shell, was typical: "There is no future for me in Shell if I can't deliver excellence in safety." The brochure for visitors to the Milliken Research and Customer Centre leaves no doubt where the responsibility lies in that company. It starts off with a quote from Roger Milliken: "Milliken believes that all injuries are preventable, all health risks are controllable, and management is accountable."

The answers to the critical question about management accountability (question 8) showed a striking difference between the safe companies and those with poor safety. In the safe companies, 75% of the respondents said that line managers were held *fully or fairly accountable* for injuries, in the companies with poor safety, only 16%.

As with management commitment, the influence of line managers and supervisors driving safety pervades all aspects of safety in the organization and shows up particularly in the attention to thorough implementation of the safety practices.

INVOLVEMENT IN SAFETY ACTIVITIES, TRAINING, AND EMPOWERMENT

The model proposes that the main vehicle for embedding safety values and developing safety awareness is extensive involvement of the entire workforce in "doing things in safety." Training is considered important, but it is slanted towards a learn–do–learn cycle. Empowerment is the natural companion of involvement.

Leading companies are increasingly turning to self-management systems to improve their performance through tapping the initiative of all their people. At DuPont Canada and Milliken, self-management systems are deeply entrenched, and Shell is moving in that direction. These systems rely heavily on involvement. For many years, workforce committees have done much of the safety work at DuPont. Milliken's emphasis on involvement is impressive. At the time of this survey, Milliken's measurement of the extent of involvement of their work-force in safety activities outside employees' regular jobs was 70% (they routinely measure involvement). To reinforce this, they do not have safety specialists.

A big difference in involvement in safety activities between the very safe companies and those with poor safety showed up in the answers to questions 9 and 16. For example, in the answers to question 16, 71% of workers in the very safe companies said that they were regularly involved or at least had some involvement in workplace audits. None of the workers in the unsafe companies were regularly involved, and only 8% said they had even some involvement.

In the answers to question 10, 87% of workers in the very safe companies said that they were fully or quite empowered to take action in safety; in the one company with poor safety where this question was asked, only 17% said they were so empowered.

In the very safe companies, training is intensive and ongoing. Many examples were found in the interviews and focus group studies. In question 11, 69% of workers in the very safe companies said that they had received extensive or considerable training in the last 2 years; in the companies with poor safety, only 8% said this.

COMPREHENSIVE SAFETY PRACTICES

The model describes an ideal in which the commitment of management, the ownership taken by line management, and the concepts of involvement are all in place and focused on the implementation of a comprehensive set of safety practices. Although each of the safe companies has its individual areas of emphasis, all stress the key practices, such as safety meetings, safety rules, and injury investigation. The questionnaire results reflected a clear difference between the quality of the practices in the very safe companies and those in the companies with poor safety. Two examples are cited below.

One of the most important practices involves safety rules—their quality, the extent to which they are followed, and the thoroughness in enforcing them. In the safe companies, 48% of the respondents said that their rules were excellent and 92% said they were good or excellent, compared with 7% and 42% in the companies with poor safety. There was an equally clear difference in the extent that the rules are obeyed and enforced.

There was a striking difference in the extent of recognition for safety achievements. In the very safe companies, 90% of the respondents said that recognition was thorough and extensive or at least frequent, compared with 22% in the companies with poor safety.

SAFETY ORGANIZATION AND SAFETY SPECIALISTS

Many of the safety practices are managed through the safety organization. In the safe companies it is operated by the line organization. Typically the leader chairs the central committee in the safe companies.

The respondents in the very safe companies gave their safety organization good marks; 88% said that it was good or excellent, compared with 40% in the companies with poor safety.

In the very safe companies, much of the safety work is done by the line organization, including workers, rather than by safety specialists. Milliken does not even have specialists on staff, a visible message of where the responsibility lies. The safety specialists were given a good rating by the respondents in the safe companies and a poor rating by those in the companies with poor safety.

SATISFACTION WITH SAFETY PERFORMANCE

The very safe companies are not as highly satisfied with their safety performance as one might expect. DuPont has very few lost work injuries and a very low total injury frequency, yet it continues to strive to eliminate those few injuries. Milliken has made great strides in improving safety but is determined to improve further. Both Milliken and Shell have the DuPont benchmarks in their sights. A focus on continuous improvement was evident in all the discussions and showed up in the answers to question 24 about satisfaction with safety. Only half of the respondents in the very safe companies said that they were very satisfied. Surprisingly, some respondents in the companies with poor safety were also *very satisfied*, and half were at least moderately satisfied. This is despite an injury rate more than 200 times worse!

VALIDITY OF THE MODEL AND THE QUESTIONNAIRE

The research results validate the model. Clearly, management commitment is essential. Line ownership of safety is really just the extension of that commitment. Workers must be knowledgeable, skilled, and safety-aware. Involvement, training, and empowerment are the best ways to instill those values, and some combination of them is needed in every workplace.

The research was not intended to differentiate among the practices, but some are obviously more important than others. Here we must fall back on experience and judgement. By and large, the package of practices hangs together as the necessary combination to help advance to world class excellence. All the practices were in place to a lesser or greater extent in all of the very safe companies. In the companies with poor safety, some practices were missing entirely.

The questionnaire is a unique tool for measuring the state of safety management. The author knows of no other work that both creates a comprehensive model and then is able to benchmark the results in a quantitative way. It yields a surprisingly accurate measure of the state of most aspects of safety management.

Some of the questions are more important than others. Likely a reasonable gauge of safety performance could be obtained with six to eight questions, for example those about the priority given to safety (1 and 2), about the preventability of injuries (3), about management accountability (8), about involvement and training (10 and 11), and about safety rules (12 and 13). However, almost all of the questions yield insights and expose deficiencies. They provide a good opportunity to discuss improvement in specific terms with the organization being assessed.

9

APPLYING THE RESULTS OF THE RESEARCH

The leaders of organizations and consultants trying to create a step change in safety find that the tools for planning and effecting the change are inadequate:

1. There are few good techniques to "measure" the state of safety management, in particular to measure the intangibles such as management commitment. Thus assessment of safety is usually observational and anecdotal rather than quantitative.
2. Although there are descriptions of how the safest companies manage safety, because of the lack of measurement tools there is little quantitative benchmark data. There is a poor understanding of what constitutes "world class safety," particularly understanding of the central role played by specific safety values.
3. Partly because of the inadequacy of the assessment tools and the lack of benchmark data, management is often reluctant to undertake the fundamental changes required to reform safety. Better ways are needed to help convince them that a step change can be managed through orderly processes.

As a result, improvement efforts often do not focus on the most important things. These are usually not the physical or system deficiencies that are the easiest to see. Rather, they are the intangibles that cannot be easily measured and thus are often neglected—things like lack of management commitment, a low level of involvement of workers in safety activities, or lack of enforcement of safety rules.

Before undertaking the research project described in this book, the author's consulting company had developed a Safety Survey process. It addresses the first of the factors—the measurement of the state of safety management. The central innovation of the Safety Survey was a questionnaire, completed by a cross section of the organization, which "measures" the state of safety management.

The research described in this book addressed the second of the above-described issues—the lack of comprehensive information, particularly quantitative benchmark data, on how the safest companies achieve excellence. The Safety Survey process, including the key tool of the process—the questionnaire—was used to benchmark the management methods of very safe and quite unsafe companies. On the other side of the coin, through the research, the model of safety management and the questionnaire were tested and improved and new insights were gained on the fundamentals of safety management.

The information and insights developed in the research can be used in a general sense by organizations striving to improve their safety. This conventional aspect of the use of the research does not need explanation. However, to understand how the research relates to the barriers to improvement cited above, the steps involved in applying the techniques will be outlined.

The Safety Survey process is the first of two related techniques developed by the author. The second process, called "Future State Visioning" (FSV) is a methodology for planning and managing change that can be used to build a plan from the survey results. It engages the leaders and others in the organization in a participative process to create a vision of future excellence in safety and to develop a path forward for improvement. The integrated process, combining the two techniques—the "Safety Improvement Process," provides a comprehensive package for managing a step change in safety.

APPLICATION OF THE SAFETY SURVEY

A typical Safety Survey would include the following steps:

1. Completion of questionnaires by a cross section of people in the organization
2. Analysis of statistical data, examination of safety manuals and other documents

3. On-site assessment of the facilities and discussions with workers and others

4. Formal focus group discussions with groups of workers and groups of supervisors

5. Interviews with the key managers and selected others

Completing the questionnaires is the first step. The subsequent steps—observation, focus groups, interviewing and data collection—can then be more focused, because they can be based on the specific shortfalls identified through the questionnaire. For example, if the questionnaire data indicated that safety rules were not being well observed, the reasons for this could be investigated in the focus groups and interviews.

The Safety Survey delivers an assessment that is based on the views of the organization itself, not just on the opinions of the consultant. It provides evidence of the fundamental changes that are needed. It identifies specific deficiencies in a way that is highly credible and points to how they can be corrected. The contrast between the company's results and the benchmarks is a natural "gap analysis" that provides a solid foundation for planning change. The quantitative aspect of the questionnaire makes it an excellent tool for tracking the progress of safety improvement through later resurveys.

Four levels of insight can be gained from the questionnaire data:

1. The absolute level of the answers yields important information. For example, if most of the answers indicate that safety rules are often not obeyed, there are serious safety problems.

2. Comparison among the answers from management, supervision, and workers yields important insights into how safety is managed and into the overall safety culture.

3. The results from the subject company can be compared with those of other companies, particularly those with excellent and those with poor safety.

4. The results obtained at one time can be compared with those obtained later to help determine whether improvement has occurred.

An example of the output from a survey at a company with mediocre safety that was striving to improve is given in Fig. 9-1.

The level of training at Company K was given a low rating by respondents in all job categories. Fewer than half of the workers said that they

The Extent of Safety Training (Company K)			
	% Who Said Their Training in the Last Two Years Has Been ...		
	Extensive or Considerable	Some	Little or None
Managers	0	80	20
Supervisors	11	42	47
Workers	15	30	55
All	12	39	49

Figure 9-1   The relatively low level of training at Company K (Q11).

Safety Training of *Workers*			
	% of *Workers* Who Said Their Training in the Last Two Years Has Been ...		
	Extensive or Considerable	Some	Little or None
BM Best Result	94	4	2
BM Safe Co. Avg.	69	24	7
Company K	15	30	55
BM Unsafe Co. Avg.	8	29	63
BM Worst Result	4	8	88

Figure 9-2   Safety training of workers at Company K was much lower than at the very safe companies (Q11).

had received any training in the previous 2 years, a result that even without benchmark comparisons indicated an important area for improvement (Fig. 9-2). When the results were compared with the benchmarks, the shortfall became even more obvious. The level of training was closer to that of the companies with poor safety than to those with excellent safety.

With these results in hand, in the focus groups and interviews, specific information was sought on what training was given and where it was deficient. The subsequent recommendations were persuasive because they were solidly based on quantitative assessments provided by the workers themselves and were backed up with specific information on the deficiencies. This example illustrates the power of the technique: such a result would be difficult to obtain with conventional observational consulting techniques.

A second example is illustrated in Fig. 9-3 for a company with very good safety that wanted to improve to world class.

Recognition for safety achievements was perceived to be quite low by all respondents, particularly by workers (Fig. 9-4). Although the results themselves were indicative of the need for change, this was underlined by the comparison with the benchmark data.

In the subsequent focus groups, the workers cited chapter and verse. They thought that the decline in recognition showed that management

Recognition for Safety Achievement (Company J)				
	% Who Said Recognition Is ...			
	Thorough, Extensive	**Frequent**	**Some**	**Little or None**
Managers	23	62	15	0
Supervisors	8	31	54	7
Workers	0	21	54	25
All	8	35	44	13

Figure 9-3   Recognition was generally low at Company J (Q19).

Workers' Views on Recognition for Safety Achievement				
	% of Workers Who Said Recognition Is ...			
	Thorough, Extensive	**Frequent**	**Some**	**Little or None**
BM Best Result	100	0	0	0
BM Safe Co. Avg.	51	37	9	3
Company J	0	21	54	25
BM Unsafe Co. Avg.	6	14	22	58
BM Worst Result	8	8	4	80

Figure 9-4   Workers felt that the level of recognition was low at Company J (Q19).

was no longer so committed to safety. Management was surprised by these results. They agreed that recognition had been reduced, partly because a falloff in safety performance had made recognition less deserved! They had not realized how this had been perceived. There were obvious solutions, and they were quickly applied.

Much of the power of the survey comes from the comparison of the results to best practices. However, the quantitative results should be brought to life with examples from companies with safety excellence and bolstered with concrete proposals for correction.

In some cases, the Safety Survey results are enough to point the way to improvement and to create the motivation for the fundamental changes in safety management that are usually needed. The survey results deliver a strong message. If the company has reasonable systems, and if the management is motivated to make a step change, the Safety Survey may be sufficient. Often, however, management does not fully appreciate the depth of reform needed nor understand how to build a plan from the results of the survey. In particular, the power of a specific vision of excellence and a clear set of beliefs may not be fully understood. In such cases, the author has found that the effectiveness of the safety improvement project can be greatly enhanced if the Safety Survey is combined with the Future State Visioning technique to provide a complete package of analysis and planning for change.

COMBINING THE SAFETY SURVEY WITH FUTURE STATE VISIONING—THE FUTURE STATE VISIONING WORKSHOP

The Future State Visioning process (Fig. 9-5) is a methodology for planning and managing change (4, 15). It is ideally suited to developing a vision of safety and articulating the beliefs (values) to underpin the vision.

It is particularly powerful when it is combined directly with the Safety Survey in an integrated two-step process. While the survey results are fresh in the minds of the leaders, this is the best time to engage them in envisaging how they want safety to be in the future. What results do they expect, what beliefs do they espouse, how do they intend to act?

The relationship between the two processes and how they help to improve the critical steps in a drive to reform safety are illustrated in Fig. 9-6.

In the integrated process, the leaders and a cross section of the orga-

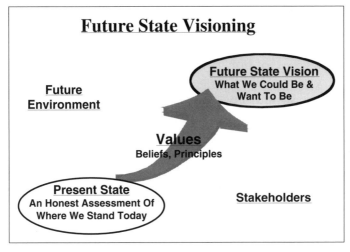

Figure 9-5 The Future State Visioning process.

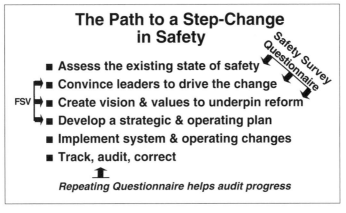

Figure 9-6 How the Safety Survey & the Future State Visioning processes interact in the critical stages of safety reform.

nization are engaged in a FSV Workshop where they plan together where they want to be in safety. A typical workshop would involve 35–50 managers, supervisors, professionals, and workers for 2–3 days. There should be strong representation of workers, particularly of opinion leaders. Their input and their later involvement in communicating and implementing the outcome are essential.

A typical workshop would start with a presentation on world class safety (how the safest companies do it), followed by presentation and discussion of the results of the Safety Survey for that company. The participants would then be engaged in team workshops (10–12 people) on the desired future safety performance, on safety leadership, on the safety attitude of workers, and on the safety practices.

In each team workshop, the goal would be to build a specific vision, *in actionable terms*, of how the organization wants safety to be in the future. The vision would then be compared with the present, using the survey results as input into "how things are today." The team would then develop the beliefs that underpin and support the desired future state. After each team workshop, the results would be discussed in plenary sessions.

Through discussion of the results in a workshop setting, with a cross section of people with different jobs and different perspectives, a much clearer understanding of the meaning of the Safety Survey results is developed. The foundation for a comprehensive, actionable plan for reform is laid. And perhaps most importantly, through involvement, those who have responsibility for implementation begin to take ownership of the process and become motivated to drive for results. The involvement of opinion leaders from all levels, particularly from the working level, helps ensure unity and buy-in of the Safety Survey results and the FSV Workshop output.

A diagram of the integrated process is shown in Fig. 9-7.

The Safety Improvement Process

The Safety Survey

- Complete questionnaires
- Compile results
- Collect data on safety
- Observe at location
- Conduct focus groups of workers & supervisors
- Interview managers & others
- Analyze results, compare to benchmarks

The Future State Visioning Workshop

- Presentation on managing for world class safety
- Presentation of Safety Survey results
- Workshops on future vision of safety & values
- Refine results of workshop -- vision & values

Follow-Up

- Communicate results
- Plan & take action
- Track progress, audit, correct

Figure 9-7 The agenda for an integrated safety improvement process.

ACTION FROM THE SURVEY AND WORKSHOP RESULTS

The purpose of the FSV Workshop is not to determine the specific action steps needed to achieve the vision, although they will be apparent from the gap analysis. Those responsible for action will have been part of the process, and each participant will come out of the workshop with specific actions that he or she can take, without waiting for an overall plan. However, it is essential that the management team develop a detailed path forward. After the FSV Workshop, the output should be condensed into a report that can be used for communication and as the basis for determining action. The team that develops the report and the action plan should include workshop participants, but management must lead it so that the vision and values become owned by those responsible for its implementation.

WHERE THE SAFETY IMPROVEMENT PROCESS HAS BEEN USED

The Safety Survey and the FSV process have been used, separately and in combination, in a variety of projects. As one would expect, they have been most successful when the company has had a genuine resolve to make a step change and needed help to define how to do it.

The FSV process was used in the early 1990s to help reform safety at National Rubber, a medium-sized Toronto company. Their 10-fold improvement in safety performance is described in Reference 4. The FSV process was also employed at Imperial Oil in a series of workshops to build a stronger joint safety and quality effort with the contract truckers that distribute their products. It was used in the city of Calgary in the mid-1990s to help establish a base for safety improvement. The FSV process was used at the Workplace Safety & Insurance Board of Ontario (WSIB) to help develop objectives and to engage the organization in building a vision and values in safety. The same techniques were used with WSIB staff and outside stakeholders to build a new policy framework for government-supported safety research in Ontario. The FSV process has been used in a variety of other cases involving safety and business planning.

The Safety Survey technique was first used in 1994 in a safety project with Abitibi-Consolidated. The Kenora Mill radically improved its safety performance, from an average of seven lost work cases a year in the early 1990s to none in the 5 years from 1995 through 1999. This will be described in a forthcoming paper (25). The Safety Survey technique was used at Bowater Mersey's Nova Scotia operations to help lay the foundation for continuing improvement in safety.

The integrated Safety Improvement Process has been used in several major projects. At Shell Canada Resources in the mid-1990s, it helped lay the foundation for an improvement in safety that brought the Resources Group back to the same excellent level as the rest of Shell Canada. It was also used at BC Gas in a comprehensive assessment of the state of safety and a process to engage leaders in change. Most recently, the integrated process has been used in major projects at Unilever Canada and at Nova Scotia Power, in each case to help the company define the route to their goal of world class safety.

The author knows of no other process that integrates measurement of the state of safety management with a process to develop a future vision and values. The quantitative measurement of the "soft" elements of safety management and their comparison with quantitative measurements at benchmark companies with world class safety is unique. No other such data are available.

FUTURE USE OF THE SAFETY IMPROVEMENT PROCESS

In 2000, the rights to the commercial use of the copyrighted elements of the Safety Survey, including the safety questionnaire, and rights to the commercial use for safety projects of the copyrighted elements of the Future State Visioning process were acquired by the E. I. DuPont de Nemours and Company. They are being incorporated into DuPont's worldwide safety consulting services.

APPENDIX A

REFERENCES AND END-NOTES

1. *Work Injuries and Diseases, Canada, 1994–1996*, Association of Workers' Compensation Boards of Canada, Mississauga, 1997. Also summary report in *OHS Canada*, Vol. 15, No. 2, March 1999.
2. *The Cost of Work-related Injury and Disease*, Australian Government Publishing Service, Canberra, 1994. A detailed analysis by Worksafe Australia estimated that occupational injuries and diseases cost their economy between $20 and $40 billion Australian per year—in an economy considerably smaller than Canada's.
3. J. M. Stewart, "The Multi-ball Juggler," *Business Quarterly*, Vol. 57, No. 4, Summer 1993.
4. J. M. Stewart, "Future State Visioning Technique (and the Turnaround in Safety) at National Rubber," *Planning Review*, Vol. 22, No. 2, March–April 1994.
5. J. M. Stewart, "Managing for World Class Safety," Report on Research on the Management of Safety, The Rotman School of Management, University of Toronto, June 1999 (published by J. M. Stewart Enterprises Inc.)
6. James R. Thomen, *Leadership in Safety Management*, John Wiley & Sons, NewYork, 1991.
7. William J. Mottel, Joseph F. Long, and David E. Morrison, *Industrial Safety is Good Business: The DuPont Story*, Van Nostrand Reinhold, New York, 1995.
8. Pascal Dennis, *Quality, Safety and Environment*, ASQC Quality Press, Milwaukee, 1997.
9. Robert D. Buzzell and Bradley T. Gale, *The PIMS Principles*, The Free Press (Macmillan), New York, 1987.

10. Harry S. Shannon, Vivienne Walters, Wayne Lewchuk, Jack Richardson, Lea Anne Moran, Ted Haines, and Dave Verma, "Workplace Organizational Correlates of Lost Time Accident Rates in Manufacturing," Ontario Workers' Compensation Institute Report, 1994.

11. Dan Petersen, "Behaviour-based Safety Systems: A Definition and Criteria to Assess," *Professional Safety*, January 1997.

12. Thomas R. Krause, "Trends and Developments in Behaviour-Based Safety," *Professional Safety*, October 1997.

13. Dan Petersen, *Safety By Objectives*, Van Nostrand Reinhold, New York, 1996.

14. Nick W. Hurst, Stephen Young, Ian Donald, Huw Gibson, and Andre Muyselaar, "Measures of Safety Management Performance and Attitudes to Safety at Major Hazard Sites," *J. Loss Prev. Ind.*, Vol. 9, No. 2, 1996.

15. J. M. Stewart, "Future State Visioning—A Powerful Leadership Process," *Long Range Planning*, Vol. 26, No. 6, 1993.

16. J. M. Stewart, "Empowering Multinational Subsidiaries," *Long Range Planning*, Vol. 28, No. 4, 1995.

17. Joe Crunk, "Using Benchmarking to Achieve Safety Excellence," presentation at OSH '94 Conference, Toronto, October 1994. Joe Crunk, Corporate Manager, Safety & Ergonomics at Intel Corporation (a former DuPont manager).

18. Assefa Bequele, "The Costs and Benefits of Protecting and Saving Lives at Work: Some Issues," *International Labour Review*, Vol. 123, No. 1, 1984.

19. "The Costs of Accidents at Work," *Health and Safety Series Booklet HS(G)96*, Health and Safety Executive Information Centre, Sheffield, U.K., 1993.

20. Thomas R. Krause and Ronald M. Finley, "Safety and Continuous Improvement," *The Safety and Health Practitioner*, September 1993.

21. Michael B. Weinstein, "Improving Behaviour-based Safety Through TQM," *Professional Safety*, January 1998.

22. Foster C. Rinfort, Jr., "Does Safety Pay?" *Professional Safety*, April 1998.

23. Woodside and Kellogg/JGC/Kaiser (KJK), Case Study by Worksafe Australia, Camperdown, NSW, Australia, 1992.

24. The Advanced Management Skills Program, begun by Walter Mahler (now deceased) in the early 1970s, continues to be offered by The Mahler Co., Inc., Fair Lawn, N.J.

25. J. M. Stewart, "Managing for World Class Health and Safety: A Turnaround in the Pulp and Paper Industry," presentation to OH&S Forum, PaperWeek International, Montreal, 1998 (to be published in Professional Safety in late 2001).

26. Phillip Crosby, quoted in Frank E. Bird, and Ray J. Davies, *Safety and the Bottom Line*, Institute Publishing, Georgia, 1996.

27. Don F. Jones, "Occupational Safety Programmes—Are They Worth It?" *Report for Labour Safety Council of Ontario*, Ontario Ministry of Labour, 1973.

28. Herbert H. Lank, *The DuPont Canada History*, DuPont Canada Inc., 1982.

29. James Chisholm and Perry Bamji, assisted by John Gordon, and J. M. Stewart, "The National Rubber Company (Parts A and B)," Business Case Study, School of Business, Queen's University, Kingston, 1996.

APPENDIX B

NOMENCLATURE

CCPA Canadian Petrochemical Producers' Association

CPPI Canadian Petroleum Products Institute

Disciplinary Action (for safety infractions) Defined as a range of actions from a cautionary conversation through to more severe action such as termination.

Injury Frequency Several injury frequencies are used here. They are all defined per 200,000 hours of work exposure, although some of the data provided to the author was in terms of injuries per 100 people (which is approximately the same).

Generally, the injury data is based on the definitions of the US agency OSHA, which are used by many Canadian companies. In some cases, such as with the CCPA member companies, long-term occupational illnesses that eventually result in lost time are included. They may not be included in some others.

- **LWIF or Lost Work Injury Frequency** An injury severe enough that the person cannot return to work on his or her next work day or shift.

- **RWIF or Restricted Work Injury Frequency** An injury where the person does not lose a day or a shift but cannot work on the regular job and is given modified duties.

- **MTIF or Medical Treatment Injury Frequency** An injury where professional treatment is required but the injury is not severe enough to result in lost time or restricted work.

- **First Aid Cases** Injuries that are treated with simple first aid.
- **OTJIF or Off-The-Job Injury Frequency** The frequency of lost work injuries that occur off the job, calculated on the individual's average time off the job, excepting sleep time.
- **TRIF or Total Recordable Injury Frequency** The total of lost work, restricted work, and medical treatment injuries.

IRS or Internal Responsibility System (Described as follows (the author's definition): "The CEO is directly responsible and accountable for the safety of all employees. In a similar way, each manager and supervisor is directly responsible and accountable for the employees in his or her jurisdiction. Each employee is directly responsible and accountable for his or her own safety and in an indirect sense for the safety of co-workers. Each of these workplace parties is fully (100%) responsible and accountable, but their responsibilities have different scope."

JH&SC or Joint Health and Safety Committee A workplace committee comprising representative of workers and management, usually mandated in legislation (sometimes JOHSC for joint occupational health and safety committee).

OH&S or Occupational Health and Safety Used to denote long- and short-term injury and occupational illness.

OSHA Occupational Health and Safety Administration, U.S. Dept. of Labor.

PIMS Profit Impact of Market Strategies Strategic Planning Institute (Reference 9).

Line Management or The Line Defined as those in the hierarchical line who do the direct work, starting with the CEO and cascading down to and including individual workers.

Values Defined as the combination of beliefs (what we hold to be true) and principles (the guidelines for translating beliefs into action).

Workforce All the people in the organization, for example, everyone in a plant from the manager to the workers.

APPENDIX C

QUESTIONS FOR INTERVIEWS OF COMPANY LEADERS
(QUESTIONS TO ASK THE CEO, OTHER SENIOR LINE MANAGERS SUCH AS PLANT MANAGERS, UNIT HEADS)

The questions were asked "cold," without "leading the witness," to get spontaneous answers. "Division," "organization," or "plant" was used in place of "company" if the respondent was other than the CEO. In some cases the unit leader was asked the same question about the *company* and about the *unit*.

VISION, VALUES, GOALS, AND OBJECTIVES

1. Does your company (unit) have a written safety vision? A set of safety values?
   1.1. How and when were they developed?
   1.2. How important are they to your safety effort?
   1.3. How are they used?
2. What are the key beliefs that your company (unit) holds about safety? (list 3 or 4)
   2.1. How important are they to your safety effort?
   2.2. What are the reasons your organization strives for excellence in safety?
3. What are the main long-term safety goals of your company (unit)?
   3.1. How and when were they developed?
   3.2. How are they used?

4. Does your company (unit) have specific (corporate) safety objectives for this year?
   4.1. What are they (quantitative as well as qualitative)?
   4.2. How were they developed? What was your role?
   4.3. Does their attainment affect your pay? Who decides?

PERSONAL ROLE AND INVOLVEMENT IN SAFETY ACTIVITIES

5. What are the most important elements of your personal role in safety? (list 3 or 4)
6. What regular duties in safety are you involved in? (Central Safety Committee?)
7. What percentage of your time is spent on safety?
8. How do you personally demonstrate your commitment to safety?
9. What role do you play when an injury or major incident occurs?
10. Are you involved in safety activities outside of your company?

LINE OWNERSHIP

11. Does the company (unit) have a top management safety committee?
    11.1. Who chairs it?
    11.2. How often does it meet?
    11.3. How do you influence the corporate (unit) safety organization?
12. Do the managers who report to you have specific safety objectives for this year?
    12.1. How were they developed?
    12.2. Does reaching them affect their performance rating and compensation?
13. Whom do you hold mainly responsible for delivering excellence in safety?
14. Whom do you hold responsible for safety training?
    14.1. Who does it?
    14.2. Does your company (unit) have training in safety values and attitude?

MANAGEMENT INFORMATION SYSTEMS, COMMUNICATIONS SYSTEMS

15. What safety information (data) do you rely on to tell you how the company (unit) is doing?

    15.1. Who prepares it?

    15.2. How often do you get it?

16. How do you assess the safety attitude in the company (unit)?

    16.1. What formal tools does your company (unit) use?

17. How does your company's (unit's) safety compare to that of others in your industry?

    17.1. To others beyond your industry?

    17.2. Who prepares the data? How often is it updated?

    17.3. How is it used?

18. How do you report your safety performance to the company (unit) at large?

    18.1. How often?

    18.2. Do you report safety performance directly to employees? How?

19. How do you report your company's (unit's) safety performance to the board? (management committee, etc., for internal corporate units).

    19.1. How often?

COST-BENEFIT BALANCE, BUSINESS-SAFETY INTERACTION

20. How do you and your company (unit) view the relationship between the management of safety and the management of other parameters?

21. At what level of safety performance do you feel that the costs of improving safety become greater than the economic benefits (cost-benefit trade-off theory)?

22. Do you think that striving for excellence in safety affects the ability of your company (unit) to be excellent in other areas—quality, costs, profits? Help or hinder?

    22.1. Could you sustain excellence in business if safety was only average?

23. How does your company (unit) deal with conflicts in priorities between safety and quality, costs, and profits?

    23.1. Have there been specific instances in the past?

24. How do you communicate the priority of safety to those that report to you?

    24.1. To the company (unit) at large?

25. Is safety included in business plans? In what way?

26. Your company has combined excellence in safety with good business results—quality, costs, productivity, and profits. How have you been able to do this?

RESOURCES—FUNDS AND PEOPLE

27. Do you have a corporate (unit) safety department?

    27.1. To whom does it report?

28. To what extent do you rely on the safety department and safety advisors to generate excellence in safety?

29. Do you have issues about funding of safety?

    29.1. How much does safety cost your company (unit) annually?

    29.2. How many full-time people are directly involved in safety?

SAFETY RULES AND DISCIPLINE

30. What role do safety rules play in your safety management?

    30.1. Who is responsible for developing and revising rules?

31. Does your company (unit) have a written policy on disciplinary action for infractions of safety rules?

    31.1. What is it?

    31.2. How and when was it developed?

    31.3. How is it used?

VARIOUS COMPANY POLICIES

32. Is safety considered in your hiring processes? How?

33. What part does safety training play in your organization?

    33.1. What is your training philosophy?

    33.2. Who is responsible for training?

34. What are your company's philosophy, policy, and practices about
    34.1. Modified duties for injured workers who cannot do their regular job?
    34.2. Rehabilitation processes for workers?
    34.3. Return to work initiatives?
    34.4. Who manages these processes?
35. What are your company's policies about safety in foreign subsidiaries?
    35.1. How and when were they developed?
    35.2. What is the experience?
36. What are your company's policies about safety of contract employees?
    36.1. How and when were they developed?
    36.2. What is the experience?
37. What is your company's (unit's) involvement in the off-the-job safety of employees?
38. What is your company's practice about the participation of employees in safety activities outside of the company?

INVOLVEMENT OF WORKFORCE

39. How involved is the workforce in general in safety activities?
40. What are your views on safety meetings?
    40.1. Is there a company or unit policy on frequency?
    40.2. How are workers involved?
41. Is there a statutory worker-management joint health and safety committee?
    41.1. What does it do?
    41.2. How effective is it?
42. How do you handle union involvement in safety?
43. How is worker involvement balanced with the role of safety advisors?
44. What are your views on safety and self-management (self-directed teams, etc.)?

RECOGNITION

45. How does your company (unit) recognize and reward safety achievements?
46. How are you personally involved?

SATISFACTION WITH PERFORMANCE

47. How satisfied are you with your company's (unit's) safety performance?

    47.1. With the rate of improvement in performance?

48. What are the most difficult barriers that had to be overcome for your company (unit) to achieve excellence in safety?

    48.1. To stay at an excellent level and continue to improve?

49. How satisfied are you with your company's (unit's) performance in quality?
50. How satisfied are you with your company's (unit's) profit performance?
51. Your company (unit) has a fine safety record. What are the main reasons (list 3 or 4)?

APPENDIX D

STATISTICAL ANALYSIS OF DATA

In the early part of the research project, the numbers of people to be surveyed in a company or company unit were chosen from practical considerations and from the author's experience in using the questionnaire in consulting. Ideally this would have been determined by prior statistical analysis. Partway through the project, some statistical analysis was conducted to guide the rest of the research.

The survey at S&C Electric was used as an example. The calculations were designed to determine whether the differences in answers between groups (e.g., between managers and supervisors or between managers and workers) were significant and whether the S&C answers differed from the best result for that question. Some examples of the statistical analyses are presented below.

The numbers of people sampled at S&C Electric are shown in Fig. AppD-1.

The S&C responses for Question 1 are summarized in Fig. AppD-2.

The statistical analysis of the internal S&C results indicated almost 95% confidence that there is a real difference between the average response of managers (82%) and that of workers (91%), more than 95% confidence that there is a difference between the response of managers (82%) and supervisors (62%), and more than 99% confidence that there is a difference between the response of supervisors (62%) and that of workers (91%).

We can be more than 99% confident that there is a real difference in results between the combined answer for all S&C respondents

Responses to Questionnaires S&C Electric		
Group	People in Company	Number of Questionnaires
Managers	11	11
Supervisors	30	13
Professionals	35	11
Workers	145	30
Other	--	--
Total	~250-300	65

Figure AppD-1   Responses to questionnaires, S&C.

The Priority Individuals Give to Safety				
	% Who Ranked Safety First			
	Manag.	Super.	Work.	All
S&C Electric	82	62	91	75
Best Result	97	90	92	94

Figure AppD-2   The priority individuals give to safety at S&C Electric (Q1).

Line Management Accountability, S&C Electric				
	% Who Said Line Management Is Held *Fully* Accountable ...			
	Managers	Supervisors	Workers	All
S&C Electric	27	47	43	37
Best Result	83	76	44	61
Safe Co. Avg.	58	55	34	44
Unsafe Co. Avg.	23	9	7	10
Worst Result	14	0	65	4

Figure AppD-3   The extent line management at S&C Electric is held *fully* responsible for safety (Q8).

(75%) and the combined answer for the company with the best result (94%).

In question 8 about the extent to which management is held account-able for safety, the S&C results had a different pattern than from that of the other very safe companies (Fig. AppD-3). The combined result for all S&C respondents was somewhat lower than the average of the very safe companies; the low result for managers pulled it down.

For the internal S&C responses, we can be more than 95% certain that there is a difference between the perception of managers (27%) and supervisors (47%) and about 95% certain that there is a difference between managers and workers (27% vs. 43%). However, the differ-ence between the responses of supervisors and workers is not signifi-cant (47% vs. 43%).

We can be confident at more than the 99% level that the S&C answers for "all" (37%) are different from those of the company with the best result (61%). Without analysis it is obvious that the S&C result for "all" respondents combined (37%) is at about the safe company average (44%) and very different from the results of the companies with poor safety or the worst result.

The statistical calculations confirmed that the sample sizes had generally been sufficient to yield satisfactory results, although larger samples would have been appropriate in some cases.

APPENDIX E

THE SAFETY QUESTIONNAIRE

SAFETY QUESTIONNAIRE

© **Original Copyright J. M. Stewart Enterprises Inc., 2000**
1 Aberfoyle Cres., PH5
Etobicoke, Ontario M8X 2X8

Copyright Assigned to E. I. DuPont de Nemours and
Company, 2000
DuPont Safety Resources
Suite 307
131 Continental Drive
Newark, DE 19713

Identification of Questionnaire

Organization: _____

Location: _____

Job Category: _____

Date Completed: _____

Reference Number: <u><This line for office use></u> _____

General Instructions & Information

This questionnaire is intended to help assess the state of safety in your organization. (Safety is used here to include both safety & occupational health.)

Your answers will be kept completely confidential. They will not be reported individually but will be combined with the answers of others. Please do not sign your name.

The person administering the questionnaire will define the organization that you should think about when answering the questions. Usually it will be the plant or location where you work. If you are uncertain of this, or of the meaning of any of the questions, please ask.

Please:
1. Answer all the questions honestly and objectively so that the answers will help reveal the true state of safety.

2. Unless otherwise indicated, do not mark more than one answer to a multiple choice question, or the answer will not be usable.

3. When asked to rank items, do not give items equal rank, even though they may seem to be close to you in their meaning. If you do, the answer will not be useable.

If you have comments, please write them on the page provided at the end of the questionnaire.

Identification of Job Category

☞ **What is your job category? Please check the appropriate category below:**

☐₁ **Management**—means the top management group or team of your organization or workplace. The leader may be the plant manager, president, CEO or other leader as defined by the person explaining the questionnaire. Management also includes managers who respond directly to that leader and others who generally could be called senior management.

☐₂ **Supervision**—includes middle management, superintendents, supervisors and foremen—those who have jobs between top management and the working level.

☐₃ **Working Level**—means operating, maintenance, clerical and non-supervisory people.

☐₄ **Professional**—means financial, technical and other specialist people that work at a professional level (such as an engineer or an accountant) but who do not supervise other people.

	Yes	No
Please also indicate whether you are involved in safety in an official capacity, for example as a safety supervisor, a safety advisor or specialist, a safety steward or a member of a joint occupational health and safety committee (JOHSC). (The normal managerial or supervisory responsibility for safety is not "official" in this sense.)	☐₁	☐₂
Please indicate whether you are a member of a union.	☐₁	☐₂

Safety Questionnaire

☞ There are 24 questions in the Questionnaire. The first two, which appear on this page, are general questions about the organization's priorities. They ask you to "Rank" several items. The remainder of the questions are "Multiple Choice," which ask you to check only one response.

1. Indicate the **priority you personally give** to the following items. Rank in order from 1 to 4, with the item you think is the most important marked 1 and the least important marked 4.

Item	Your Priority
A. Quality, customer focus	
B. Costs, efficiency	
C. Production volume	
D. Safety	

2. Indicate where **you think others** in your organization rank the same items.

   For example, give your opinion of the priority that you think supervision as a group gives to the item. Rank in order of priority from 1 to 4 as in question 1.

Item	Priority of Management	Priority of Supervision	Priority of Workers
A. Quality, customer focus			
B. Costs, efficiency			
C. Production volume			
D. Safety			

3. To what extent can **injuries be prevented**?

Check the answer that represents your personal belief.

☐₁ All can be prevented

☐₂ Almost all can be prevented

☐₃ Many can be prevented

☐₄ Some can be prevented

☐₅ Few can be prevented

4. Indicate how you think that a drive (a strong, long term effort) for excellence in safety would affect **the ability to achieve excellence in other areas**, such as quality, productivity, costs and profits. Check only one answer.

The safety effort will:

☐₁ Be very helpful in achieving other business objectives.

☐₂ Provide some positive assistance in reaching other business objectives.

☐₃ Have neither a positive nor a negative effect on attaining other business objectives.

☐₄ Tend to make it more difficult to achieve other business objectives.

☐₅ Substantially weaken the ability to achieve other business objectives.

5. At what level of safety performance do you think that the effort to improve safety starts to **cost more than it yields in "economic benefits"**? Economic benefits means savings from the reduction in the costs of injuries, lost working time, loss of material, etc., and the indirect economic benefits that good safety brings in better morale, improved production, better product quality, etc. Check only one answer.

☐₁ Within reason there is no limit. *Exceptional* safety performance returns more in economic benefits than it costs.

☐₂ Once safety performance is at the *excellent* level, further improvement will cost more than the economic benefits it delivers.

$\square_3$ Once *good* (well above average) safety performance is reached, further improvement will cost more than the economic benefits it delivers.

$\square_4$ Once *average* safety performance is reached, further improvement will cost more than the economic benefits it delivers.

$\square_5$ Safety programs are a net cost. Improving safety always costs more than it benefits.

6. To what extent is safety "designed in" to the facilities and **"built in"** to the operating practices in your organization? Built in means that safety is considered as an integral part of the design of equipment, of the development of operating practices and of job training, not something that is added on later. Check only one answer.

$\square_1$ Thoroughly built in

$\square_2$ Substantially built in

$\square_3$ Some integration

$\square_4$ Little integration; mainly added later

$\square_5$ Not integrated at all; added later

7. Does your organization have well-established, written **safety values**—sometimes called beliefs and principles? "Written" means readily available in a document, posted on the bulletin board, etc. Check only one answer.

$\square_1$ Yes

$\square_2$ No

$\square_3$ I don't know

If you answered "No" or "I don't know," please go to question 8.

If you answered "Yes," indicating that your organization does have written safety values, please check the statement below that best describes those values:

$\square_1$ We have safety values and they are up-to-date, well understood and have an important influence on safety.

$\square_2$ We have safety values and they have some influence on safety.

$\square_3$ We have safety values but they are not used much and they have little influence on safety.

8. Indicate the extent to which **line managers are held accountable** for injuries and safety incidents in their areas. (Line managers include such titles as supervisor, foreman, superintendent, team leader, etc., as well as manager.) Check only one answer. In our organization:

☐₁ Line managers are held fully accountable for preventing injuries and incidents in their area. Safety performance has a direct effect on their performance rating and pay. This is a key part of our safety management.

☐₂ Line managers are held accountable for preventing injuries and incidents in their area but safety performance does not generally affect their performance rating and pay.

☐₃ Line managers are held accountable for injuries and safety incidents but only in a general way.

☐₄ While line managers take some responsibility for injuries and incidents in their areas, most injuries are attributed to individual error, bad luck or unfortunate circumstances.

☐₅ Injuries and incidents are almost always blamed on individual error, bad luck or unfortunate circumstances. Safety is much less important than business factors, such as costs and profits, in the assessment of managers' performance.

9. How actively have you been **involved** in safety activities **in the last year**? Involvement means not just attending meetings but participation in doing things in safety such as being on a committee, participating in an investigation, or helping put together safety rules. Check only one answer.

☐₁ Deeply involved

☐₂ Quite involved

☐₃ Moderately involved

☐₄ Not much involved

☐₅ Not involved at all

In the last **two years**, have you either been on a standing safety committee (for example a JOHSC, a rules and procedures committee, a safe driving committee, etc.) *or* on a specific task force or team (for example, a team formed to review the safety rules of an area)?

☐₁ Yes   ☐₂ No

10. To what extent are you **empowered to take action** to ensure your own safety and that of others with whom you work? *Empowered* means that you are expected to take whatever action is required to avoid injuries to yourself or to others, including shutting down equipment. It means that you are empowered to fix unsafe situations within the scope of your job, expected to take whatever action is required in urgent cases and encouraged (expected) to make recommendations even if the situation does not relate to your job. It means that your organization has created an environment in which everyone is encouraged to contribute to improvement. It means that you have had safety training and involvement in safety activities that give you confidence to act.

On the other hand, *not empowered* means that for safety matters, you are expected to stick pretty strictly to the confines of your specific job description. Check only one answer.

☐$_1$ Fully empowered

☐$_2$ Quite empowered

☐$_3$ Moderately empowered

☐$_4$ Not very empowered

☐$_5$ Not at all empowered

11. Indicate the extent to which you have received **training** in safety and occupational health **in the last two years**. Training includes formal training courses away from the job and organized on-the-job training. Check only one answer.

☐$_1$ Thorough & extensive training

☐$_2$ Considerable training

☐$_3$ Some training

☐$_4$ Little training

☐$_5$ No training

12. Are **safety meetings** held regularly in your workplace? If so, how often?

☐$_1$ Every week or every two weeks

☐$_2$ Every month

$\square_3$ Every two months

$\square_4$ Less frequently than every two months

$\square_5$ Not regularly held

Do you attend the safety meetings regularly?

$\square_1$ Yes $\square_2$ No

How do you rate the **quality and effectiveness** of the safety meetings? Consider how well attended they are. Consider the content of the meetings and the extent of involvement of people in developing and conducting them.

$\square_1$ Excellent

$\square_2$ Good

$\square_3$ Satisfactory

$\square_4$ Poor

$\square_5$ Very poor

13. Please consider the **quality of the safety rules** in your organization. High quality rules are up-to-date and clearly written. They are well understood by those doing the work and help them to do the job well and safely. Check only one answer.

The quality of our safety rules is:

$\square_1$ Excellent

$\square_2$ Good

$\square_3$ Satisfactory

$\square_4$ Poor

$\square_5$ Very poor

To what extent are the **safety rules** of your organization **obeyed**? Check only one answer.

$\square_1$ All safety rules are obeyed without exception.

$\square_2$ People generally obey the safety rules.

$\square_3$ The safety rules are guidelines, sometimes followed, sometimes not.

☐$_4$ The safety rules are often not obeyed.

☐$_5$ People pay little attention to the safety rules.

14. Indicate your opinion of the way that **disciplinary action** is used in your organization for infractions to safety rules or practices. "Infraction" means breaking a safety rule or not following a standard practice. Disciplinary action refers to the range of actions, from a cautionary conversation or warning through to more severe action such as termination. Check the one answer you believe is the most accurate.

☐$_1$ Disciplinary action, related to the seriousness of the infraction, is taken for all safety infractions.

☐$_2$ Disciplinary action is taken only for serious safety infractions.

☐$_3$ Disciplinary action for safety infractions is applied arbitrarily and inconsistently.

☐$_4$ Disciplinary action is seldom taken for safety infractions.

15. To what extent are injuries, safety incidents, near misses and the like in your organization **investigated**, reported and action taken? Check only one answer.

☐$_1$ All injuries and incidents are thoroughly investigated and the recommendations implemented.

☐$_2$ Most injuries and incidents are investigated and most of the recommendations are implemented.

☐$_3$ Many of the injuries and incidents are investigated and some of the recommendations are implemented.

☐$_4$ Only the most serious injuries and incidents are investigated.

☐$_5$ Injuries and incidents are not usually investigated.

16. Indicate the extent that you are **personally involved in safety audits** and inspections of the workplace. Involvement means participation on a regularly scheduled, organized basis, not informal, personal inspections. Check only one answer.

☐$_1$ Regularly involved

☐$_2$ Some involvement

☐$_3$ Not involved at all

How do you rate the **quality and effectiveness** of the safety audit and inspection system? Consider the frequency and thoroughness of the inspections, the extent of participation of the workforce, the extent to which safety behaviour is observed, as well as physical conditions, the thoroughness of the follow-up and the overall effectiveness in helping develop a safer workplace. Check only one answer.

☐₁ Excellent

☐₂ Good

☐₃ Satisfactory

☐₄ Poor

☐₅ Very poor

17. Good management of workplace safety includes strong efforts to find temporary **modified duties** for injured people who cannot do their regular job but who can safely do other work. When people are off work because of injuries, effective action is taken to assist in their rehabilitation and to ensure their early **return to work**. Line management and supervision take responsibility for these efforts and they are conducted in a thorough but sympathetic manner. In this context, please rate the effectiveness of the modified duty and return-to-work initiatives of your organization.

☐₁ Excellent

☐₂ Good

☐₃ Satisfactory

☐₄ Poor

☐₅ Very poor

18. To what extent is **"off-the-job" safety** dealt with in the workplace safety program of your organization? Check only one answer.

☐₁ Off-the-job safety is an important part of our safety program. We keep statistics on off-the-job injuries, have an off-the-job safety committee, programs to promote safety in the home, safe driving off-the-job, etc.

☐₂ Off-the-job safety is not a formal part of our workplace safety program but aspects of it are sometimes included in safety meetings, etc.

☐₃ Off-the-job safety is not part of our workplace safety program.

19. Indicate the extent to which achievements in safety are **recognized** and good safety performance is celebrated in your organization. Check the one answer that best represents your opinion.

☐₁ Thorough and extensive

☐₂ Frequent

☐₃ Some

☐₄ Little

☐₅ None

20. How do you rate the safety of the **physical facilities** in your workplace (machinery, equipment, etc.)? Check the one answer that describes your assessment.

☐₁ Excellent

☐₂ Good

☐₃ Satisfactory

☐₄ Poor

☐₅ Very poor

21. Check the one answer that best represents the extent of your **personal knowledge of the safety performance** of your organization:

☐₁ I know our safety goals and our up-to-date performance. I know how our performance compares with that of other companies in our industry.

☐₂ I know our safety goals and present performance but do not know how our performance compares to that of other companies.

☐₃ I am only generally aware of our safety goals and how we are doing in safety. I do not know how we compare to others.

$\square_4$ I do not know our safety goals. I am not familiar with how we are performing in safety. I do not know how we compare to others.

22. How do you rate the effectiveness of the **safety organization** in your workplace (the managers' safety committee, the JOHSC, other safety committees, the safety systems, structures and procedures, etc.)? Check only one answer.

$\square_1$ Excellent

$\square_2$ Good

$\square_3$ Satisfactory

$\square_4$ Poor

$\square_5$ Very poor

23. How do you rate the effectiveness of the **safety department** in your organization (the safety supervisor, the safety advisors, safety specialists, etc.)? Check only one answer.

$\square_1$ Excellent

$\square_2$ Good

$\square_3$ Satisfactory

$\square_4$ Poor

$\square_5$ Very poor

24. To what extent are you personally **satisfied with the safety performance** of your organization? Check only one answer.

$\square_1$ Very satisfied

$\square_2$ Moderately satisfied

$\square_3$ Neither satisfied nor dissatisfied

$\square_4$ Moderately dissatisfied

$\square_5$ Very dissatisfied

COMMENTS

COMMENTS

_____

_____

_____

_____

_____

_____

_____

_____

_____

_____

_____

_____

_____

_____

_____

_____

_____

_____

_____

_____

_____

_____

_____

_____

_____

_____

_____

© **Original Copyright J. M. Stewart Enterprises Inc., 2000**
1 Aberfoyle Cres., PH5
Etobicoke, Ontario M8X 2X8

Copyright Assigned to E. I. DuPont de Nemours and Company, 2000
DuPont Safety Resources
Suite 307
131 Continental Drive
Newark, DE 19713

APPENDIX F

TABLES OF DETAILED RESULTS

The sources of the questionnaire data for the tables are summarized below.

NOTES ON THE TABLES

(The data is essentially as presented in the research report (5) with some minor corrections and revisions. For four questions a different "worst result" than in the research report was deemed to be more appropriate.)

1. The tables that follow include the detailed data, numbered as in the questionnaire. Questions 1 and 2 involved ranking four factors. For these questions, the detailed data referring to *safety* are given, but only summary data are given for the other factors. Otherwise all of the data is included.

2. In most cases, the job category of the individual could be determined from the information on the completed questionnaire. In the few cases where the job category could not be identified, the individual results were included in the database as "undesignated." Thus the number of respondents listed under "all" is sometimes greater than the total of the numbers shown for managers, supervisors, professionals, and workers. For simplicity, the undesignated results were not listed separately.

Number & Distribution of Questionnaires

Job Category	Five Very Safe Companies	Five Companies With Very Poor Safety	Total
Corporate Officers	25	0	25
Other Managers	73	29	102
Total Managers	98	29	127
Supervisors	65	73	138
Professionals	41	14	55
Workers	191	116	307
Undesignated	4	21	25
Total Valid Data	399	253	652
Rejected	3	15	18
Total	402	268	670
Average LWIF, 1993-7	0.08	20	~

Figure AppF-1   Sources of data for the tales that follow.

3. The responses from corporate management were included with those from plant management in one grouping called managers or management.

4. Some companies asked that the detailed data for their companies not be revealed separately. Thus the data are given as averages for the five very safe companies and for the five companies with very poor safety. The identity of the companies with the best and worst results is not given. For the same reason, the numbers of respondents is not given for the best or worst result.

5. The best or worst result was not necessarily for the same company for all parts of multipart questions.

6. In some cases, the answers to one or more questions from a questionnaire were not useable but the rest were. If five or more were unusable, the whole questionnaire was discarded. Some questions were added or changed as the research proceeded. Thus there were not the same number of respondents for each question. The number of companies surveyed (if less than the standard five) and the number of answers included in the averages are usually indicated at the bottom of the table.

7. Most of the data are expressed as the percentage of a given group that gave the indicated answer. In questions 1 and 2, the respondents were asked to rank the factors in order of priority from 1 to 4. For these questions an "average ranking number" from 1 (first priority) to 4 (fourth priority) was calculated. For example, if nine of thirteen managers ranked the particular factor first, one ranked it second, and three ranked it fourth, the average ranking number would be 1.7. The average ranking number gives a rough indication of the overall ranking or priority given to a particular factor. The number can also reveal additional information, for example, if no respondents ranked an item first but all ranked it second. In these questions, respondents were required not to give any of the factors equal ranking or the answer was discarded.

8. In question 2, the column "selves" is the average of the answers given by individuals in question 1, i.e., their own individual priority. The last column in the table is the *average* of the columns for managers, supervisors, and workers. (The respondents were not asked to rate the priority of the professional group.) In all of the other tables, the corresponding column is called "all" and is the average of all the individual answers. It is a weighted number, because each individual answer has equal weight; and thus the largest group—workers—has the greatest influence on the result.

9. The numbers called "average, very safe companies" and "average, companies with very poor safety" were not weighted. They are the simple averages of the average results for the five companies, regardless of the number of respondents.

10. In question 7, Table (c), the percentages refer to all of the respondents. Those who answered "no" or "don't know" when asked about the presence of safety values (Table (a)) were inferred to believe that the values were "of little influence." Some respondents ignored the instructions that only those that answered

"yes" to the first part of the question should proceed to the second part. When more or fewer answered part 2 than should have, the results were prorated on the basis of the answers of those that did answer. This usually resulted in only a small correction.

11. "Safe Co." and "Unsafe Co." are used in the figures in the chapter as short forms for the very safe companies and the companies with very poor safety, respectively. "Unsafe companies," "companies with very poor safety," and "companies with poor safety" are also used interchangeably, as are "safe companies," "companies with very good safety," and "companies with good safety," etc.

Q1: The Priority Individuals Give to Safety
Table (a)

Priority To Safety	Best Result					Average, Very Safe Companies					Average, Companies With Very Poor Safety					Worst Result				
	Man	Sup	Prof	Work	All	Man	Sup	Prof	Work	All	Man	Sup	Prof	Work	All	Man	Sup	Prof	Work	All
1st	97	90	100	92	94	84	79	71	91	83	46	64	30	65	62	44	58	60	53	56
2nd	3	10	0	5	4	11	14	17	7	11	28	20	14	19	22	44	16	20	32	25
3rd	0	0	0	3	2	3	7	8	2	5	22	13	47	12	12	12	15	0	15	12
4th	0	0	0	0	0	2	0	4	0	1	4	3	9	4	4	0	11	20	0	7
1st and 2nd Combined	100	100	100	97	98	95	93	88	98	94	74	84	44	84	84	88	74	80	85	81
3rd and 4th Combined	0	0	0	3	2	5	7	12	2	6	26	16	56	16	16	12	26	20	15	19
No. of Respondents	~	~	~	~	~	97	63	41	177	382	29	69	15	110	237	~	~	~	~	~

Table (b)

| Factor | Average Ranking Number Given By Group* | | | | | | | | | | | | | | | | | | | |
| | Best Result | | | | | Average, Very Safe Companies | | | | | Average, Companies With Very Poor Safety | | | | | Worst Result | | | | |
	Man	Sup	Prof	Work	All	Man	Sup	Prof	Work	All	Man	Sup	Prof	Work	All	Man	Sup	Prof	Work	All
Safety	1.0	1.1	1.0	1.1	1.0	1.2	1.3	1.4	1.1	1.2	1.8	1.6	2.3	1.5	1.6	1.7	1.8	1.8	1.6	1.6
Quality	2.0	1.9	2.0	2.0	2.0	2.4	2.2	2.1	2.3	2.3	1.6	2.3	2.2	2.1	2.1	1.6	2.2	2.0	1.9	1.9
Costs	2.9	3.0	3.0	3.1	3.0	2.7	3.0	2.9	3.0	2.9	2.8	2.8	2.4	3.2	3.0	3.0	3.1	2.6	3.3	3.0
Volume	3.9	4.0	4.0	3.5	3.6	3.6	3.5	3.5	3.5	3.5	3.7	3.4	3.1	3.2	3.3	3.8	2.9	3.6	3.2	3.1

Table (c)

| Factor | % Responding Group That Gives Factor First Priority | | | | | | | | | | | | | | | | | | | |
| | Best Result | | | | | Average, Very Safe Companies | | | | | Average, Companies With Very Poor Safety | | | | | Worst Result | | | | |
	Man	Sup	Prof	Work	All	Man	Sup	Prof	Work	All	Man	Sup	Prof	Work	All	Man	Sup	Prof	Work	All
Safety	97	90	100	92	94	84	79	71	91	83	46	64	30	65	62	44	58	60	53	56
Quality	0	10	0	5	4	7	14	20	4	10	47	18	39	24	25	56	26	20	26	29
Costs	3	0	0	3	2	5	1	3	4	3	3	8	31	3	6	0	5	20	9	7
Volume	0	0	0	0	0	4	6	6	1	4	4	10	0	8	7	0	11	0	12	8

*The average ranking number was calculated by adding the ranking given by respondents in that category and dividing by the number of respondents, e.g., if all respondents ranked safety first, the ranking number would be 1.0. If all ranked volume fourth, the ranking number would be 4.0.

Q2: The Priority Respondents Think Others Give to Safety

Table (a)

% Of Responding Group Who Believed They Or Others Rank Safety First

Group Offering Opinion	Best Result					Average, Very Safe Companies					Average, Companies With Very Poor Safety					Worst Result				
	Selves	Man	Sup	Work	Avg	Selves	Man	Sup	Work	Avg	Selves	Man	Sup	Work	Avg	Selves	Man	Sup	Work	Avg
Managers	82	91	82	91	88	84	85	78	79	80	46	31	21	36	29	67	33	33	0	22
Supervisors	62	77	62	100	79	79	72	71	83	75	64	30	42	42	38	58	25	50	23	33
Workers	91	91	82	91	88	91	68	64	91	74	65	18	19	57	31	53	29	24	29	27
Professionals	55	55	55	55	55	71	45	48	78	57	30	27	20	33	27	0	0	0	0	0
All	75	81	72	86	80	83	72	69	85	75	62	23	28	49	33	56	28	35	22	28
	Man	Sup	Prof	Work	All	Man	Sup	Prof	Work	All	Man	Sup	Prof	Work	All	Man	Sup	Prof	Work	All
No. of Respondents	~	~	~	~	~	95	64	41	172	376	28	67	14	102	237	~	~	~	~	~

Table (b)

Average Ranking Number Given By Group

Group Offering Opinion	Best Result					Average, Very Safe Companies					Average, Companies With Very Poor Safety					Worst Result				
	Selves	Man	Sup	Work	Avg	Selves	Man	Sup	Work	Avg	Selves	Man	Sup	Work	Avg	Selves	Man	Sup	Work	Avg
Managers	1.3	1.1	1.2	1.1	1.1	1.2	1.2	1.3	1.3	1.3	1.8	2.3	2.4	2.0	2.2	1.7	2.3	2.3	2.7	2.4
Supervisors	1.5	1.3	1.5	1.0	1.3	1.3	1.4	1.3	1.2	1.3	1.6	2.5	1.9	1.9	2.1	1.5	2.5	1.7	1.9	2.0
Workers	1.1	1.1	1.3	1.1	1.2	1.1	1.5	1.6	1.1	1.4	1.5	3.1	2.7	1.8	2.5	1.8	2.6	2.4	2.2	2.4
Professionals	1.9	1.7	1.6	1.5	1.6	1.4	1.9	1.7	1.3	1.6	2.3	2.8	2.4	2.3	2.5	2.5	2.5	2.5	3.5	2.8

Table (c)

| Factor | Average Ranking Number Workers Say Others Give To Factors | | | | | | | | | | | | | | | | | | | |
| | Best Result | | | | | Average, Very Safe Companies | | | | | Average, Companies With Very Poor Safety | | | | | Worst Result | | | | |
	Selves	Man	Sup	Work	Avg	Selves	Man	Sup	Work	Avg	Selves	Man	Sup	Work	Avg	Selves	Man	Sup	Work	Avg
Safety	1.1	1.1	1.3	1.1	1.2	1.1	1.5	1.6	1.1	1.4	1.5	3.1	2.7	1.8	2.5	1.8	2.6	2.4	2.2	2.4
Quality	2.0	2.3	2.2	2.5	2.3	2.3	2.6	2.7	2.5	2.6	2.1	2.5	2.4	2.4	2.4	1.7	2.2	1.9	1.8	2.0
Costs	3.2	3.0	3.2	3.3	3.2	3.0	2.6	2.9	3.3	2.9	3.2	2.3	2.9	3.4	2.8	3.1	2.5	3.3	3.6	3.1
Volume	3.7	3.6	3.2	3.1	3.3	3.5	3.2	2.7	3.0	3.0	3.2	2.2	2.0	2.4	2.2	3.4	2.6	2.4	2.4	2.5

Table (d)

| Factor | Average Ranking Number Managers Say Others Give To Factors | | | | | | | | | | | | | | | | | | | |
| | Best Result | | | | | Average, Very Safe Companies | | | | | Average, Companies With Very Poor Safety | | | | | Worst Result | | | | |
	Selves	Man	Sup	Work	Avg	Selves	Man	Sup	Work	Avg	Selves	Man	Sup	Work	Avg	Selves	Man	Sup	Work	Avg
Safety	1.3	1.1	1.2	1.1	1.1	1.3	1.2	1.3	1.3	1.3	1.8	2.2	2.4	2.1	2.2	1.7	2.3	2.3	2.7	2.4
Quality	2.1	1.9	2.4	2.5	2.2	2.5	2.7	3.1	3.1	3.0	1.5	1.6	2.2	2.3	2.1	1.8	1.8	2.3	2.0	2.1
Costs	3.0	3.2	3.4	3.7	3.4	2.7	2.8	3.1	3.3	3.1	2.8	2.6	2.9	3.8	3.1	2.7	2.5	3.3	3.7	3.2
Volume	3.6	3.8	3.1	2.7	3.2	3.5	3.2	2.5	2.3	2.7	3.9	3.6	2.4	1.7	2.6	3.8	3.3	2.0	1.7	2.3

Note 1: The column "Selves" gives the answers to question 1 of individuals' priority ranking for the particular company. However, the Best Result and the Worst Result for question 2 were for different companies than for question 1.

Note 2: The "Average" column is the *average* of the percentage numbers for managers, supervisors, and workers as a group. In most of the rest of the tables, the corresponding column is called "all" and is the average of all of the individual answers.

Q3: The Belief That Injuries can be Prevented

Extent Injuries Preventable	% Of Responding Group Answering As Indicated																			
	Best Result					Average, Very Safe Companies					Average, Companies With Very Poor Safety					Worst Result				
	Man	Sup	Prof	Work	All	Man	Sup	Prof	Work	All	Man	Sup	Prof	Work	All	Man	Sup	Prof	Work	All
All	86	88	73	64	75	67	73	57	46	57	25	23	29	15	20	0	0	0	11	9
Almost All	14	12	27	36	25	33	25	40	42	37	45	57	63	57	54	83	50	100	58	58
Many	0	0	0	0	0	0	2	3	10	5	30	18	8	24	24	17	50	0	26	31
Some	0	0	0	0	0	0	0	0	2	1	0	2	0	4	2	0	0	0	5	2
Few	0	0	0	0	0	0	0	0	0	0	0	0	0	0	0	0	0	0	0	0
No. of Respondents	~	~	~	~	~	98	65	40	191	398	29	73	15	115	249	~	~	~	~	~

Q4: The Effect of a Drive for Safety on Business Performance

Effect Of Safety Drive	% Of Responding Group Answering As Indicated																			
	Best Result					Average, Very Safe Companies					Average, Companies With Very Poor Safety					Worst Result				
	Man	Sup	Prof	Work	All	Man	Sup	Prof	Work	All	Man	Sup	Prof	Work	All	Man	Sup	Prof	Work	All
Very Helpful	86	80	57	68	73	76	71	57	62	67	56	32	43	24	32	45	39	40	17	24
Positive	11	20	43	32	26	23	27	39	31	29	36	55	42	43	40	33	33	60	58	42
Neither + Nor –	0	0	0	0	0	0	2	2	0	1	6	7	15	24	15	11	14	0	17	12
Makes More Difficult	3	0	0	0	1	1	0	2	7	3	2	6	0	6	8	11	14	0	8	15
Substantially Weakens	0	0	0	0	0	0	0	0	0	0	0	0	0	3	5	0	0	0	0	7
No. of Respondents	~	~	~	~	~	98	65	41	190	398	29	73	15	113	247	~	~	~	~	~

Note: The Worst Result included data from several respondents who did not designate their job category. The "All" figure of 7% is correct even though none of the respondents who listed their job category answered "Substantially Weakens."

Q5: The Level of Safety Where the Cost-Benefit Break Point Occurs

Where Safety Starts To Cost	% Of Responding Group Answering As Indicated																			
	Best Result					Average, Very Safe Companies					Average, Companies With Very Poor Safety					Worst Result				
	Man	Sup	Prof	Work	All	Man	Sup	Prof	Work	All	Man	Sup	Prof	Work	All	Man	Sup	Prof	Work	All
No Limit	97	90	86	91	92	76	80	61	81	77	57	80	78	69	67	0	67	100	45	47
At Excellent Safety	3	0	14	9	7	24	18	31	18	21	39	3	11	13	12	100	0	0	21	23
At Good Safety	0	10	0	0	1	0	2	8	0	2	4	5	0	7	7	0	0	0	13	10
At Average Safety	0	0	0	0	0	0	0	0	0	0	0	10	5	6	5	0	33	0	8	10
Safety Always A Net Cost	0	0	0	0	0	0	0	0	1	0	0	2	6	5	9	0	0	0	13	10
No. of Respondents	~	~	~	~	~	98	65	41	190	398	22	59	13	97	204	~	~	~	~	~

Note: Data for only 4 companies with very poor safety.

Q6: The Extent to Which Safety is Built in

Extent Safety Built In	Best Result					Average, Very Safe Companies					Average, Companies With Very Poor Safety					Worst Result				
	Man	Sup	Prof	Work	All	Man	Sup	Prof	Work	All	Man	Sup	Prof	Work	All	Man	Sup	Prof	Work	All
Thoroughly	36	38	55	67	54	42	33	43	47	42	50	0	0	13	13	50	0	0	13	13
Substantially	64	62	45	33	46	55	54	38	44	47	0	100	100	4	17	0	100	100	4	17
Some Integration	0	0	0	0	0	3	10	15	7	8	50	0	0	42	36	50	0	0	42	36
Little Integration	0	0	0	0	0	0	3	0	2	2	0	0	0	33	27	0	0	0	33	27
No Integration	0	0	0	0	0	0	0	4	0	1	0	0	0	8	7	0	0	0	8	7
No. of Respondents	~	~	~	~	~	98	65	40	191	398	~	~	~	~	~	~	~	~	~	~

% Of Responding Group Answering As Indicated

Note: Data for only 1 company with very poor safety.

Q7: The Presence and Influence of Safety Values

Table (a)

Have Values	% Of Responding Group Answering As Indicated																			
	Best Result					Average, Very Safe Companies					Average, Companies With Very Poor Safety					Worst Result				
	Man	Sup	Prof	Work	All	Man	Sup	Prof	Work	All	Man	Sup	Prof	Work	All	Man	Sup	Prof	Work	All
Yes	97	100	100	100	99	89	85	83	96	90	64	56	41	52	50	60	24	67	44	38
No	0	0	0	0	0	9	11	13	1	7	17	27	24	32	31	40	58	33	56	53
Don't Know	3	0	0	0	1	2	4	4	3	3	19	17	35	16	19	0	18	0	0	9
No. of Respondents	?	?	?	?	?	91	63	38	182	378	29	72	14	110	240	?	?	?	?	?

Table (b)

Quality Of Safety Values	% Of Those Who Said "Yes There Are Values" Who Answered As Indicated																			
	Best Result					Average, Very Safe Companies					Average, Companies With Very Poor Safety					Worst Result				
	Man	Sup	Prof	Work	All	Man	Sup	Prof	Work	All	Man	Sup	Prof	Work	All	Man	Sup	Prof	Work	All
Up-to-date and Influential	82	100	71	93	89	89	67	44	82	75	21	27	0	26	24	0	25	0	0	8
Some Influence	15	0	29	7	10	10	31	53	18	24	68	61	67	55	56	67	50	50	100	69
Little Influence	3	0	0	0	1	1	2	3	0	1	11	12	33	19	20	33	25	50	0	23
No. of Respondents	?	?	?	?	?	85	55	33	175	348	17	34	9	60	123	?	?	?	?	?

Table (c)

Quality Of Safety Values	% Of Responding Group Answering As Indicated																			
	Best Result					Average, Very Safe Companies					Average, Companies With Very Poor Safety					Worst Result				
	Man	Sup	Prof	Work	All	Man	Sup	Prof	Work	All	Man	Sup	Prof	Work	All	Man	Sup	Prof	Work	All
Up-to-date and Influential	80	100	71	93	88	79	57	37	79	67	14	15	0	14	12	0	6	0	0	3
Some Influence	14	0	29	7	10	9	27	45	17	22	45	35	29	28	28	40	12	33	44	26
Little Influence	6	0	0	0	2	12	16	18	4	11	41	50	71	58	60	60	82	67	56	71
No. of Respondents	~	~	~	~	~	91	63	38	182	378	29	72	14	110	240	~	~	~	~	~

Note: The percentages in Table c) refer to all respondents. Those who answered "no" or "don't know" re the presence of values were inferred to believe that the values were of little influence. See point 10 in "Notes on the Tables" for explanation of the calculations.

Q8: The Extent to Which Line Management Held Accountable for Safety

Extent Held Accountable	Best Result					Average, Very Safe Companies					Average, Companies With Very Poor Safety					Worst Result				
% Of Responding Group Citing Indicated Accountability																				
	Man	Sup	Prof	Work	All	Man	Sup	Prof	Work	All	Man	Sup	Prof	Work	All	Man	Sup	Prof	Work	All
Fully Accountable	83	76	46	44	61	58	55	40	34	44	23	9	0	7	10	14	6	0	0	4
Held Fairly Accountable	17	18	36	24	22	35	31	29	30	31	0	13	5	1	6	0	0	25	0	6
Only Generally	0	6	9	14	8	4	7	16	17	11	54	53	75	40	47	72	76	50	54	60
Little Accountability	0	0	9	18	9	2	6	12	18	12	0	7	0	3	3	0	0	0	0	0
Not Accountable	0	0	0	0	0	1	1	3	1	2	23	18	20	49	34	14	18	25	46	30
No. of Respondents	~	~	~	~	~	98	65	40	187	394	28	73	15	114	246	~	~	~	~	~

Q9: Involvement in Safety Activities

Table (a)

Extent Involved	Best Result					Average, Very Safe Companies					Average, Companies With Very Poor Safety					Worst Result				
	Man	Sup	Prof	Work	All	Man	Sup	Prof	Work	All	Man	Sup	Prof	Work	All	Man	Sup	Prof	Work	All
Deeply	66	30	29	39	46	41	29	22	22	28	7	6	25	5	8	0	0	0	4	3
Quite	23	50	14	29	28	34	40	18	20	28	47	25	12	14	21	50	34	0	8	13
Moderately	11	10	43	20	18	13	19	17	23	20	15	35	24	16	21	0	33	100	0	7
Not Very Much	0	10	14	7	5	7	8	23	18	13	23	25	16	20	21	50	33	0	29	30
Not At All	0	~	0	5	3	5	4	20	17	11	8	9	23	45	29	0	0	0	59	47
No. of Respondents	~					98	65	41	190	398	29	73	15	116	250	~				

Table (b)

Whether Involved in Specific Tasks in Last 2 Years	Best Result					Average, Very Safe Companies					Average, Companies With Very Poor Safety					Worst Result				
	Man	Sup	Prof	Work	All	Man	Sup	Prof	Work	All	Man	Sup	Prof	Work	All	Man	Sup	Prof	Work	All
Yes	83	70	71	71	74	59	62	52	39	47	50	100	0	8	10	50	0	0	8	10
No	17	30	29	29	26	41	38	48	61	53	50	100	100	92	90	50	100	100	92	90
No. of Respondents	~					52	45	30	172	299	~					~				

Note: For part 2 of question 9 (Table (b)), there were data for only 4 very safe companies and for only 1 company with very poor safety. There were data for all 10 companies for part 1.

Q10: The Extent to Which Individuals Feel Empowered to Take Action in Safety

Extent of Empowerment	% Of Responding Group Answering As Indicated																			
	Best Result					Average, Very Safe Companies					Average, Companies With Very Poor Safety					Worst Result				
	Man	Sup	Prof	Work	All	Man	Sup	Prof	Work	All	Man	Sup	Prof	Work	All	Man	Sup	Prof	Work	All
Fully Empowered	94	90	100	81	86	82	86	54	67	73	50	67	0	17	23	50	67	0	17	23
Quite Empowered	6	10	0	13	10	16	10	34	20	19	50	33	100	0	10	50	33	100	0	10
Moderately Empowered	0	0	0	3	2	2	4	7	12	7	0	0	0	25	20	0	0	0	25	20
Not Very Empowered	0	0	0	3	2	0	0	5	1	1	0	0	0	37	30	0	0	0	37	30
Not At All Empowered	0	0	0	0	0	0	0	0	0	0	0	0	0	21	17	0	0	0	21	17
No. of Respondents	~	~	~	~	~	98	65	41	191	399	~	~	~	~	~	~	~	~	~	~

Note: Data for only one company with very poor safety.

Q11: The Extent of Safety Training

Extent Of Training	% Of Responding Group Citing Indicated Extent Of Training																			
	Best Result					Average, Very Safe Companies					Average, Companies With Very Poor Safety					Worst Result				
	Man	Sup	Prof	Work	All	Man	Sup	Prof	Work	All	Man	Sup	Prof	Work	All	Man	Sup	Prof	Work	All
Extensive Training	51	50	29	71	61	12	15	12	26	19	3	8	0	1	3	0	34	0	0	3
Considerable Training	43	20	71	23	31	37	32	44	43	38	4	14	21	7	9	0	33	0	4	7
Some Training	6	20	0	4	5	36	32	21	24	28	36	46	35	29	36	50	33	0	8	13
Little Training	0	10	0	1	2	10	18	17	6	12	36	23	5	25	25	50	0	0	29	27
No Training	0	0	0	1	1	5	3	6	1	3	21	9	39	38	27	0	0	100	59	50
No. of Respondents	~	~	~	~	~	98	65	41	191	399	29	73	15	116	250	~	~	~	~	~

Q12: The Frequency and Quality of Safety Meetings
Table (a)

Frequency of Safety Meetings	% of Responding Group Giving Indicated Rating																			
	Best Result					Average, Very Safe Companies					Average, Companies With Very Poor Safety					Worst Result				
	Man	Sup	Prof	Work	All	Man	Sup	Prof	Work	All	Man	Sup	Prof	Work	All	Man	Sup	Prof	Work	All
Every Week or Every Two weeks	91	100	86	99	96	45	59	41	57	54	0	0	0	5	4	0	0	0	5	4
Every Month	9	0	14	0	3	55	38	27	31	35	100	67	0	24	33	100	67	0	24	33
Every Two Month	0	0	0	0	0	0	3	10	4	3	0	33	0	5	7	0	33	0	5	7
Less Than Every Two Month	0	0	0	1	1	0	0	19	6	6	0	0	0	0	0	0	0	0	0	0
Not Regularly	0	0	0	0	0	0	0	3	2	2	0	0	100	66	56	0	0	100	66	56
No. of Respondents	~	~	~	~	~	45	30	23	150	248	~	~	~	~	~	~	~	~	~	~

Table (b)

Attendance At Safety Meetings	% Of Responding Group Giving Indicated Rating																			
	Best Result					Average, Very Safe Companies					Average, Companies With Very Poor Safety					Worst Result				
	Man	Sup	Prof	Work	All	Man	Sup	Prof	Work	All	Man	Sup	Prof	Work	All	Man	Sup	Prof	Work	All
Yes	0	100	80	96	95	76	79	72	74	77	50	0	0	10	12	50	0	0	10	12
No	0	0	20	4	5	24	21	28	26	23	50	100	100	90	88	50	100	100	90	88
No. of Respondents	~	~	~	~	~	45	30	23	150	248	~	~	~	~	~	~	~	~	~	~

Table (c)

Quality and Effectiveness Of Safety Meetings	% Of Responding Group Giving Indicated Rating																			
	Best Result					Average, Very Safe Companies					Average, Companies With Very Poor Safety					Worst Result				
	Man	Sup	Prof	Work	All	Man	Sup	Prof	Work	All	Man	Sup	Prof	Work	All	Man	Sup	Prof	Work	All
Excellent	47	30	14	40	40	46	40	20	35	36	0	0	0	6	4	0	0	0	6	4
Good	41	70	72	48	49	48	57	59	48	49	50	67	0	11	21	50	67	0	11	21
Satisfactory	12	0	14	12	11	6	0	21	17	14	50	33	100	22	29	50	33	100	22	29
Poor	0	0	0	0	0	0	3	0	0	1	0	0	0	61	46	0	0	0	61	46
Very Poor	0	0	0	0	0	0	0	0	0	0	0	0	0	0	0	0	0	0	0	0
No. of Respondents	~	~	~	~	~	45	29	23	149	246	~	~	~	~	~	~	~	~	~	~

Note: Data for only 3 very safe companies and for only 1 company with very poor safety.

Q13: The Quality of Safety Rules and Extent Obeyed

Table (a)

Quality Of Safety Rules	% Of Responding Group Citing Indicated Quality Of Rules																			
	Best Result					Average, Very Safe Companies					Average, Companies With Very Poor Safety					Worst Result				
	Man	Sup	Prof	Work	All	Man	Sup	Prof	Work	All	Man	Sup	Prof	Work	All	Man	Sup	Prof	Work	All
Excellent	68	60	29	67	65	50	40	33	53	48	0	3	0	12	7	0	0	0	9	3
Good	29	40	71	30	33	47	48	50	43	44	40	48	42	29	35	60	33	67	18	36
Satisfactory	3	0	0	3	2	3	9	17	4	7	53	37	44	33	38	20	39	33	27	32
Poor	0	0	0	0	0	0	1	0	0	0	7	11	14	23	18	20	22	0	37	24
Very Poor	0	0	0	0	0	0	2	0	0	1	0	1	0	3	2	0	6	0	9	5
No. of Respondents	~	~	~	~	~	97	65	41	191	398	29	73	15	116	250	~	~	~	~	~

Table (b)

Extent Rules Are Obeyed	% Of Responding Group Answering As Indicated																			
	Best Result					Average, Very Safe Companies					Average, Companies With Very Poor Safety					Worst Result				
	Man	Sup	Prof	Work	All	Man	Sup	Prof	Work	All	Man	Sup	Prof	Work	All	Man	Sup	Prof	Work	All
No Exceptions	0	8	36	37	25	6	6	17	25	16	0	0	0	4	3	0	0	0	0	0
Generally Obeyed	100	77	64	60	70	90	84	70	64	74	73	45	64	35	41	60	39	34	27	38
Sometimes Followed	0	15	0	0	3	4	10	6	10	8	17	45	29	54	44	40	44	33	64	48
Not Often Obeyed	0	0	0	3	2	0	0	7	1	2	10	10	7	5	11	0	17	33	9	14
Little Attention	0	0	0	0	0	0	0	0	0	0	0	0	0	2	1	0	0	0	0	0
No. of Respondents	~	~	~	~	~	97	65	41	191	398	29	73	15	116	250	~	~	~	~	~

Q14: The Extent Safety Rules are Enforced

Extent Disciplinary Action Taken	% Of Responding Group Answering As Indicated																			
	Best Result					Average, Very Safe Companies					Average, Companies With Very Poor Safety					Worst Result				
	Man	Sup	Prof	Work	All	Man	Sup	Prof	Work	All	Man	Sup	Prof	Work	All	Man	Sup	Prof	Work	All
All Infractions	62	80	57	70	67	46	41	36	53	45	24	24	8	18	18	14	0	0	4	4
Only Serious Infractions	26	10	29	17	20	35	24	28	22	27	24	23	35	18	21	29	18	0	12	16
Inconsistent	9	10	14	13	12	8	11	9	10	11	25	24	42	20	27	14	18	25	4	11
Seldom Taken	3	0	0	0	1	11	24	27	15	17	27	29	15	44	34	43	64	75	80	69
No. of Respondents	~	~	~	~	~	97	65	41	191	398	28	73	15	116	249	~	~	~	~	~

Q15: Thoroughness in Investigation of Injuries and Incidents

Thoroughness Of Investigation	% Of Responding Group Answering As Indicated																			
	Best Result					Average, Very Safe Companies					Average, Companies With Very Poor Safety					Worst Result				
	Man	Sup	Prof	Work	All	Man	Sup	Prof	Work	All	Man	Sup	Prof	Work	All	Man	Sup	Prof	Work	All
All Investigated	77	80	86	92	85	70	58	68	70	68	15	26	12	16	17	0	67	0	13	17
Almost All Investigated	20	20	14	6	12	27	39	29	23	27	48	40	54	32	35	50	33	100	13	20
Many Investigated	0	0	0	1	1	2	3	3	5	4	24	25	17	22	25	50	0	0	21	20
Only Serious Incidents	3	0	0	1	2	1	0	0	2	1	13	9	17	21	18	0	0	0	36	30
Not Usually Done	0	0	0	0	0	0	0	0	0	0	0	0	0	8	5	0	0	0	17	13
No. of Respondents	~	~	~	~	~	98	65	40	191	398	29	73	15	116	249	~	~	~	~	~

Q16: Involvement in and Quality of Safety Audits

Table (a)

| Extent Of Involvement | % Of Responding Group Citing Indicated Extent Of Involvement | | | | | | | | | | | | | | | | | | | |
| | Best Result | | | | | Average, Very Safe Companies | | | | | Average, Companies With Very Poor Safety | | | | | Worst Result | | | | |
	Man	Sup	Prof	Work	All	Man	Sup	Prof	Work	All	Man	Sup	Prof	Work	All	Man	Sup	Prof	Work	All
Regularly Involved	82	70	57	62	68	59	44	22	35	40	50	33	0	0	7	50	33	0	0	7
Some Involvement	12	20	43	34	27	19	29	25	36	31	50	0	0	8	10	50	0	0	8	10
Not At All	6	10	0	4	5	22	27	53	29	29	0	67	100	92	83	0	67	100	92	83
No. of Respondents	~	~	~	~	~	97	65	40	190	396	~	~	~	~	~	~	~	~	~	~

Table (b)

| Quality Of Safety Audits | % Of Responding Group Answering As Indicated | | | | | | | | | | | | | | | | | | | |
| | Best Result | | | | | Average, Very Safe Companies | | | | | Average, Companies With Very Poor Safety | | | | | Worst Result | | | | |
	Man	Sup	Prof	Work	All	Man	Sup	Prof	Work	All	Man	Sup	Prof	Work	All	Man	Sup	Prof	Work	All
Excellent	55	8	46	54	43	39	17	30	31	29	0	0	0	4	3	0	0	0	4	3
Good	45	92	36	33	48	47	57	37	57	51	50	33	0	9	14	50	33	0	9	14
Satisfactory	0	0	18	13	9	14	24	29	11	18	50	67	0	26	31	50	67	0	26	31
Poor	0	0	0	0	0	0	2	0	1	1	0	0	100	44	38	0	0	100	44	38
Very Poor	0	0	0	0	0	0	0	4	0	1	0	0	0	17	14	0	0	0	17	14
No. of Respondents	~	~	~	~	~	98	65	41	190	398	~	~	~	~	~	~	~	~	~	~

Note: Data for only one company with very poor safety.

Q17: Rating of the Modified Duty and Return-To-Work Systems

Rating, Modified Duty & Return-To-Work	% Of Responding Group Giving Indicated Rating																			
	Best Result					Average, Very Safe Companies					Average, Companies With Very Poor Safety					Worst Result				
	Man	Sup	Prof	Work	All	Man	Sup	Prof	Work	All	Man	Sup	Prof	Work	All	Man	Sup	Prof	Work	All
Excellent	64	77	82	67	70	67	56	66	48	53	50	34	0	8	13	50	34	0	8	13
Good	36	23	9	33	28	31	26	24	33	29	50	33	0	17	20	50	33	0	17	20
Satisfactory	0	0	9	0	2	2	18	10	10	11	0	33	100	21	23	0	33	100	21	23
Poor	0	0	0	0	0	0	0	0	8	6	0	0	0	46	37	0	0	0	46	37
Very Poor	0	0	0	0	0	0	0	0	1	1	0	0	0	8	7	0	0	0	8	7
No. of Respondents	~	~	~	~	~	46	30	23	152	251	~	~	~	~	~	~	~	~	~	~

Note: Data for only 3 very safe companies and for only 1 company with very poor safety.

Q18: The Presence of Off-The-Job Safety Programs

Status Of Off-The-Job Safety Program	% Of Responding Group Answering As Indicated																			
	Best Result					Average, Very Safe Companies					Average, Companies With Very Poor Safety					Worst Result				
	Man	Sup	Prof	Work	All	Man	Sup	Prof	Work	All	Man	Sup	Prof	Work	All	Man	Sup	Prof	Work	All
Integral Component	100	100	100	98	99	58	43	52	51	52	7	5	0	7	10	0	0	0	8	7
Informal Component	0	0	0	0	0	35	48	48	46	43	50	41	71	23	35	50	0	100	0	7
Not Included	0	0	0	2	1	7	9	0	3	5	43	54	29	70	55	50	100	0	92	86
No. of Respondents	~	~	~	~	~	97	65	41	191	398	29	72	15	114	245	~	~	~	~	~

Q19: Recognition for Safety Achievements

Extent Of Recognition	% Of Responding Group Citing Indicated Extent Of Recognition																			
	Best Result					Average, Very Safe Companies					Average, Companies With Very Poor Safety					Worst Result				
	Man	Sup	Prof	Work	All	Man	Sup	Prof	Work	All	Man	Sup	Prof	Work	All	Man	Sup	Prof	Work	All
Thorough & Extensive	90	80	100	100	92	63	50	59	51	55	4	1	13	6	7	0	7	0	11	7
Frequent	10	20	0	0	8	31	41	31	37	35	11	27	10	14	15	0	0	0	6	7
Some	0	0	0	0	0	6	8	10	9	9	50	25	26	22	27	0	14	0	17	14
Little	0	0	0	0	0	0	1	0	2	1	19	19	37	22	22	33	21	50	28	26
None	0	0	0	0	0	0	0	0	1	0	16	28	14	36	29	67	58	50	38	46
No. of Respondents	~	~	~	~	~	98	65	41	191	399	29	72	15	115	246	~	~	~	~	~

Q20: Rating of the Safety of Facilities and Equipment

Rating Of Safety Of Facilities & Equipment	% Of Responding Group Giving Indicated Rating																			
	Best Result					Average, Very Safe Companies					Average, Companies With Very Poor Safety					Worst Result				
	Man	Sup	Prof	Work	All	Man	Sup	Prof	Work	All	Man	Sup	Prof	Work	All	Man	Sup	Prof	Work	All
Excellent	73	62	73	60	65	59	47	39	34	44	4	5	0	6	5	0	0	0	0	0
Good	27	38	27	33	32	40	43	57	56	48	65	40	83	38	46	50	0	100	17	20
Satisfactory	0	0	0	7	3	1	10	0	10	7	24	44	17	33	31	50	100	0	33	40
Poor	0	0	0	0	0	0	0	0	0	0	7	9	0	14	13	0	0	0	37	30
Very Poor	0	0	0	0	0	0	0	4	0	1	0	2	0	9	5	0	0	0	13	10
No. of Respondents	~	~	~	~	~	98	65	41	191	399	29	73	15	116	250	~	~	~	~	~

Q21: Knowledge of Safety Performance

Extent Of Knowledge	% Of Responding Group Citing Indicated Knowledge																			
	Best Result					Average, Very Safe Companies					Average, Companies With Very Poor Safety					Worst Result				
	Man	Sup	Prof	Work	All	Man	Sup	Prof	Work	All	Man	Sup	Prof	Work	All	Man	Sup	Prof	Work	All
Full Knowledge	90	100	100	82	92	91	83	80	58	75	33	12	21	6	13	50	0	0	8	10
Know Own Company	10	0	0	18	8	9	16	20	35	22	56	48	22	35	40	50	67	0	17	23
Only General Knowledge	0	0	0	0	0	0	1	0	6	3	8	35	49	47	38	0	33	100	46	44
No Knowledge	0	0	0	0	0	0	0	0	1	0	3	5	8	12	9	0	0	0	29	23
No. of Respondents	~	~	~	~	~	98	65	41	191	399	29	73	15	115	249	~	~	~	~	~

Q22: Rating of the Safety Organization

Rating Of The Safety Organization	% Of Responding Group Giving Indicated Rating																			
	Best Result					Average, Very Safe Companies					Average, Companies With Very Poor Safety					Worst Result				
	Man	Sup	Prof	Work	All	Man	Sup	Prof	Work	All	Man	Sup	Prof	Work	All	Man	Sup	Prof	Work	All
Excellent	64	15	64	70	57	39	25	29	41	35	5	2	0	6	5	0	0	0	0	0
Good	36	85	36	20	38	53	64	56	47	53	39	40	61	34	35	0	17	67	46	27
Satisfactory	0	0	0	10	5	8	10	12	12	11	42	48	30	37	39	80	49	33	18	43
Poor	0	0	0	0	0	0	1	3	0	1	14	8	5	17	17	20	28	0	18	22
Very Poor	0	0	0	0	0	0	0	0	0	0	0	2	4	6	4	0	6	0	18	8
No. of Respondents	~	~	~	~	~	98	65	41	191	399	28	73	15	116	249	~	~	~	~	~

Q23: Rating of the Safety Department

Rating Of The Safety Department	% Of Responding Group Giving Indicated Rating																			
	Best Result					Average, Very Safe Companies					Average, Companies With Very Poor Safety					Worst Result				
	Man	Sup	Prof	Work	All	Man	Sup	Prof	Work	All	Man	Sup	Prof	Work	All	Man	Sup	Prof	Work	All
Excellent	58	50	43	39	45	39	25	27	36	33	16	10	4	8	9	0	0	0	0	0
Good	36	50	43	60	52	47	50	53	51	49	36	39	55	28	33	0	6	34	9	8
Satisfactory	6	0	14	1	3	14	23	16	11	16	48	29	17	38	36	100	39	33	18	40
Poor	0	0	0	0	0	0	2	0	2	1	0	19	20	18	17	0	44	33	46	38
Very Poor	0	0	0	0	0	0	0	4	0	1	0	3	4	8	5	0	11	0	27	14
No. of Respondents	~	~	~	~	~	96	65	41	191	397	28	73	15	116	248	~	~	~	~	~

Q24: Satisfaction with the Safety Performance of the Organization

Level Of Satisfaction	% Of Responding Group Citing Indicated Level Of Satisfaction																			
	Best Result					Average, Very Safe Companies					Average, Companies With Very Poor Safety					Worst Result				
	Man	Sup	Prof	Work	All	Man	Sup	Prof	Work	All	Man	Sup	Prof	Work	All	Man	Sup	Prof	Work	All
Very Satisfied	91	62	91	77	78	54	42	58	60	53	10	4	4	16	10	0	0	0	13	10
Moderately Satisfied	9	38	9	23	22	39	56	42	38	43	50	66	41	30	40	50	67	0	0	10
Not Satisfied or Dissatisfied	0	0	0	0	0	3	1	0	1	2	14	15	39	25	21	0	33	100	25	27
Moderately Dissatisfied	0	0	0	0	0	3	1	0	1	2	26	13	4	18	21	50	0	0	41	36
Very Dissatisfied	0	0	0	0	0	1	0	0	0	0	0	2	12	11	8	0	0	0	21	17
No. of Respondents	~	~	~	~	~	98	65	41	191	399	29	73	15	116	249	~	~	~	~	~

APPENDIX G

ABOUT THE AUTHOR

J. M. (Jim) Stewart is president of J. M. Stewart Enterprises Inc., a consulting company in strategic management and management of occupational health and safety.

Dr. Stewart retired in 1991 from DuPont Canada, where he was a senior vice president. His 30-year career there ranged across corporate and business management, planning, manufacturing, technical management, and research. Early in his career, he was manager of the DuPont Canada Research Centre. For several years he was responsible for the Company's manufacturing and engineering, as well as for a number of businesses, during a period of major change. He was one of the leaders in developing and implementing world class manufacturing, self-management, and empowerment. Throughout his career, Dr. Stewart has been involved in management of safety. When he was responsible for DuPont Canada's manufacturing, particular attention to safety was essential as the company downsized and moved to self-managing processes. In 1989, he was the first recipient of DuPont Canada's highest award, the Daedalus Award, given for his outstanding contribution to building self-management, increasing productivity, and reducing costs in the company.

From 1991 through 1998, Dr. Stewart was an Executive-in-Residence and Adjunct Professor of Strategic Management at the Rotman School of Management, University of Toronto. He was involved in research on business strategy, self-management, and safety management. While there, he undertook the research project funded by the federal government, provincial governments, and corporations that is the subject of this book.

Through his consulting and research, Dr. Stewart has developed unique techniques in safety management. The "Safety Survey" process develops quantitative measurements of the state of safety management. The "Future State Visioning" technique engages leaders in a process of understanding where they are in safety and in helping them to build a foundation for a future change. Linked together, these techniques provide a powerful methodology for reforming safety. Their development and application have established Dr. Stewart as a leading researcher and consultant in safety management. In 2000, the copyrights to the Safety Survey and Future State Visioning processes were acquired by E. I. du Pont de Nemours, Wilmington, DE, USA, and are being applied in their safety consulting business.

He has been active in the Canadian Society for Chemical Engineering (President), the Canadian Manufacturers' Association (Ontario Chairman & National Board), The Industrial Accident Prevention Association (Advisory Board), Minerva Canada (Board of Directors), and other associations. From 1996 to 1998, Dr. Stewart was a Director of the Ontario Workplace Safety and Insurance Board (previously the WCB) and was Chair of its Health and Safety Committee.

Dr. Stewart is a chemical engineer with a B.A. Sc. from Toronto and a Ph.D. from Imperial College, University of London. He is married with two children and two grandchildren. Among his hobbies is collecting antique waterfowl decoys, and he has co-authored a book on the subject.

INDEX